iPadOS/iOS 15

User's Guide

For Senior Citizens

Comprehensive Guide to Hidden Features, Tips And Tricks of the New Apple iOS 15 and iPadOS 15

Paul

Spurgeon

Copyright

Printed in the United States of America
© 2021 by Paul Spurgeon

Table of Contents

SECTION THREE

SECTION FOUR

SECTION ONE

Introduction

Apple has released the iOS/iPad 15 and these operating systems are said to be huge improvements to previous ones. For instance in iPadOS, widgets now rest on the Home screen, instead of being relegated to the sidebar. Also we can get new Contacts and Find My widgets in the iPadOS 15. Just like in previous iPadOS, App Library will also be present in iPadOS 15 thereby making it accessible through the device's Dock. Now in this version you can hide Home screens completely as well as remove apps you might not need frequently. New widget layouts for the Home Screen and App Library feature uncomplicated means to classify apps and customize iPad's experience. The Home Screen in iPadOS/iOS 15 is designed with integrated widgets and App Library, FaceTime calls, new equipment to minimize distractions, a redesigned Safari display, Quick Note and much more. With iPadOS 15 new insightful multitasking experience are Split View and Slide Over are straightforward to locate, simple to use and well robust. The Notes app appears system wide with Quick Note, and delivers new means to combine and organize whether writing or typing with Apple Pencil. The Translate app offers new features for translating text and conversations, and with Swift Playgrounds you can now design apps for iPhone and iPad on your iPad. Apple iOS/iPad 15 comes with some tools to minimize distractions, a new notifications experience, new features for FaceTime calls, extra privacy features redesigned Weather, Maps, and Safari and so on. With iOS/iPad 15 FaceTime calls look more natural, and it includes SharePlay for sharing experiences, helps you focus and be in the moment with new processes to group notifications, and intelligence to images and search to easily access information. Apple Maps flashes an awesome new approach to study the world. The Weather app was not left behind, as it is restructured with full-display maps and extra graphical interface of data. The Wallet app comes with support for home keys and ID cards, and surfing the internet with Safari is even easier with a new tab bar design and Tab Groups. Also, these updates come with new privacy controls in Siri, Mail, and several other interfaces across the operating system to give additional protection to user data.

What is New in iPadOS|iOS 15?

Home Screen Pages

With iPadOS/iOS 15 you can seamlessly reorder Home Screen pages, and even hide pages to streamline the Home Screen. With the new Focus mode, you can also make certain Home Screen pages come up conditionally.

Widgets

This feature was introduced into the iPadOS in iPadOS 14, however was fixed to the left side of the Home Screen. And it was not so with iPhones and this design makes the iPad widgets difficult to access.

Now with iPadOS or iOS 15 running in your device you can place widgets anywhere you desire on the Home screen or even make them dwell among apps on your device's home screen. Thereby featuring more data at a glance and delivering a more customized experience.

In addition we now have new, huge widget options, different sizes and widgets for more apps when compared to the iPadOS/iOS 14. The new widgets include that for Find My, Mail, Contacts, Game Center, and App Store. We now have some huge widgets and they are designed for iPad with huge displays for displaying pictures, music, games, videos and much more.

App Library

All credit to Apple's team of engineers, we now have an App Library on iPadOS and it is even present on the Dock of iPadOS 15. This feature automatically assembles apps into insightful categories such as Games, Sports, Productivity, Recently Added, and frequently used. Thereby making it easy for people to access apps from the Dock. Also you can hide and reorder app pages too, just like on the iPhone. In a simpler approach, App Library is a menu where you can place every apps in your device in a separate window the same pattern as the App drawer on Android. While in iPhones you will have to swipe to the end of the home screen to access the App Library but in your iPad you can effortlessly launch the app from the Dock from any window you are in.

Multitasking

In iPadOS 15, multitasking was redesigned and at the top of each app window there is a new multitasking menu (in the form of three dots), which enables you to navigate to Slide Over or Split View easily by a click.

Instead of dragging or dropping apps around like in the past, just click this new menu at the top of each app's window to go to Slide Over or Spilt View. Users can speedily access the Home Screen now while on Split View, making it easy to switch to the right apps. With the brand new shelf, multitasking with apps is so easy with multiple windows such as Safari and Pages. This enables users to work seamlessly with multiple apps as well as swiftly preview emails. This shelf displays all opened windows at the bottom of your device and you can pull apps down to the shelf by swiping down. Also, apps can be placed against each other in the multitasking view to generate new split views.

Keyboard

We now have new keyboard shortcuts to access every new multitasking feature in iOS and iPadOS 15 and you can easily set up and navigate between Split View and Slide Over with new shortcuts of the keyboard. You can now choose text correctly via an enhanced cursor that magnifies text.

You can enter Vietnamese using VNI and VIQR.

In these iOS/iPadOS updates we have new languages that support QuickPath such as Swedish, Russian, Turkish, Hindi (Latin), Vietnamese, and Dutch.

Also the updates have new keyboard layouts for Fula (Adlam), Igbo, Amharic, Ainu, Syriac, Navajo, Tigrinya, and Rohingya.

Alongside numerous additions to keyboard features is the Enhanced 10-key layout for Chinese Pinyin. The Enhanced 10-key layout encompasses easy access to more symbols, development that allows you to quickly swap to QWERTY, and type words that share the same keys with higher accuracy by letting users choose the exact Pinyin for more than just the first syllable in the phrase.

Users will also get Dialect lexicon support for Cantonese and Shanghainese. So you can enter words in Pinyin via native Shanghainese or Cantonese dialectal spellings.

Thanks to Apple Engineers, Smart Replies now support 10 new Indic languages, and they include Kannada, Bangla, Tamil, Punjabi, Urdu, Marathi, Gujarati, Odiya, Telugu and Malayalam.

With iOS and iPadOS 15, users can explore the built-in dictation in numerous languages and regions, which includes Japanese, English, French, Yue Chinese, German, Italian, Korean, Mandarin Chinese, Russian, Spanish, Turkish, Arabic, and Cantonese.

Alongside the on-device dictation, you can dictate text of any length without a timeout. In previous iOS and iPadOS dictation was previously limited to 60 seconds.

Quick Note

This is a new feature in iPadOS 15 that allows you to swiftly and easily take notes in any window across the device. You can pop up the Quick Note from any menu to write down an idea and bring in links, creating an open channel to return to the exact place you were working in. The Notes app has new Tags, a Tag Browser, and tag-based Smart Folders. Shared notes now have mentions to prompt collaborators and an Activity view. Tags help us to classify notes and access them easily with an all-new Tag Browser and tag-based Smart Folders. When using shared notes with other people, mentions gives a means to prompt collaborators and connect them back to the note. And a new Activity view displays latest updates in the note. The Notes app also comes with a new approach to collaborate, unite and grab information. You can tag people into documents, access a revision history and much more. Additional new enhancement is the capacity to easily launch a Quick Note by swiping upward on the screen with the Apple Pencil, offering a channel to seamlessly sketch a note or put down your thoughts. An Activity view allows the user to check on all of the changes made to a note while you were away. And we now have now #tags, making it easier to unite everything present in a note. You can access a note from anywhere by swiping up from the bottom right of the screen with your stylus. Quick Note can bring in data from an app in your note, if you have already opened the app. And the Notes app uses intelligence to determine this. Quick Notes will only be present on Mac and iPadOS. Quick Note can be minimized and open links in the background.

Focus

This is a brand new feature in iPadOS/ iOS 15 that helps to filter notifications and apps depending on your preference – that is what you use. If you set up Focus on one of your Apple devices, it automatically applies to the rest of your devices. When a user's Focus is hindering incoming alerts, their status will be shown to others in Messages. With iPadOS and iOS 15 Focus will be suggested for distinct situations, like winding down for bed or work hours using on-device intelligence, but you can also design a custom Focus. This new Focus feature helps you to focus and minimize distractions. You can as well design Home Screen pages with widgets and apps that apply to moments of focus to only display useful apps and prevent distractions. In a nutshell, the Focus is created to filter out notifications when you need to pay attention to a certain work. For instance, you might want to work and don't want to get WhatsApp messages. So you can set up your device via the Focus to filter out unwanted notifications when working.

Live Text

This new feature in iPadOS and iOS 15 makes use of built-in intelligence to identify text in a photo and lets you make a move. For instance, a snapshot of the front of a car showroom may pop up phone numbers with the option to place a call. With Visual Look Up, you can recognize objects in a picture, like the kind of flower or dog breed. This feature also works with handwritten text — excellent when looking for images of notes or whiteboards.

Spotlight

Via this feature users can now look for images by people, objects, location, text and scenes in iOS/iPadOS 15. It also features web photo search and excellent results for TV shows, movies, actors, and musicians. Upgraded output for contacts display recent conversations, shared photos, and location when it is shared using Find My. Capitalizing Live Text, Spotlight gives an easy means to see a public transit map, screenshot of a cooking recipe or receipt.

Translate

With iPadOS and iOS 15, this app has new features, which gives a more natural and easy means of conversations. While using the app, Auto Translate recognizes a person is speaking, and the language. 2 that is, you can talk naturally without having to click a microphone button. During 1personal conversations, face-to-face view let two individuals to sit across from each other with an iPad between them and view translations of their conversation from each own side. With iPadOS 15/iOS 15 text can be translated on windown by highlighting and clicking Translate. And the handwritten text can be translated as well. In integration with Live Text, you can also translate text in pictures.

FaceTime

With iOS/ iPadOS 15, FaceTime calls have Spatial Audio and voice isolation that makes voices sound as if they are approaching from the individual on the display. User can now come up with shareable links to a scheduled FaceTime call and can also be launched on other devices like Windows and Android. Also, FaceTime is compatible with Portrait mode and delivers a new grid view to view numerous faces at the same time. FaceTime enables you to easily share links with the relevant people, and interactions with friends and family looks more natural.

SharePlay

This is a new functionality that enables you to share media together in sync during a FaceTime call? We can now share experiences with SharePlay when connected with family or pal on FaceTime. Through this feature you can view movies or TV shows from Apple TV and other streaming platforms in sync, sharing their display to view apps or listening to songs jointly with Apple Music. Shared playback controls allow you to play, pause, or jump ahead. Also it allows people to share their display, and is great when ruminating the web as one. You can now stretch playback to Apple TV, and view it on your TV when sharing screen with the loved ones on FaceTime. However this feature has been discontinued in this update, we hope Apple will come out with better innovation in the not too distant future.

Safari

In iPadOS/ iOS 15 the popular Safari has a new-fashioned tab bar design that emits the color of webpages and brings together tabs, the toolbar, and the search field into a compact design. There are also tabs categorized for organizing tabs more easily within your device and web extensions for the first time. This new design in Safari, lets you view additional pages while browsing. The Tab Groups option allows you to easily save and organize tabs, which is ideal when you need to arrange shopping, trips or just saving frequently used tabs. Tab Groups will sync across all your devices, including iPhone and Mac. Safari is compatible with extensions in the App Store.

Swift Playgrounds

Swift Playgrounds is the simplest and great means to learn how to code. This is an app that gives you not just the platform to learn how to code but also to build apps right on your iPad for both iPhones and iPads. And you publish the apps directly on the App Store. When building an app, code is immediately reflected in the live preview and you can run the app full screen for trial. These Playgrounds make use of a new open project format based on Swift packages, which can be launched and be edited within Xcode on the Mac, delivering more versatility to developers to develop apps across Mac and iPad. Generally it enables you to build iPad and iPhone apps via SwiftUI. With iPadOS 15, there is excellent code completion, and you can as well publish the iOS and iPadOS apps directly from the Swift Playgrounds app, without ever having to use a Mac.

Notifications

In iPadOS/iOS 15 notifications were utterly redesigned. Your alerts will now pop up huge icons for apps and contact images for human beings for easy recognition. We now have an all-new customized notification summary, which combines non-urgent notifications to be sent at the due time, like in the evening afternoon or morning. Via on-device intelligence that studies your communication with apps, notifications in the summary are organized in the order of precedence with most relevant and important alerts coming first. However, emergency messages and time-band alerts will be delivered speedily. With iPadOS/iOS 15 you now have the capacity to mute any app or messaging thread's alerts momentarily. And these updates will advocate muting a thread when it is abnormally active but you are not interacting with it. We also have a new notification API designed for developers, letting them send time-band notifications and embrace the new look for alerts surfacing from individuals.

Universal Control

This new feature of iPadOS 15 lets you make use of a keyboard, trackpad or mouse across the Mac and the iPad. While migrating to iPad from Mac, the cursor for the trackpad or mouse converts to a round dot from an arrow and switches configuration automatically to suit the device. Setup process is not required. Users can easily move the cursor from one device to another until it shows up on the other device. And you can switch your cursor effortlessly between the two. You decide to set up Universal Control to function regularly in the System Preferences menu, hindering the need to connect the two devices every time. You can drag and drop items back and forth between devices, offering a better platform to draw with Apple Pencil on iPad and placing it into a Keynote slide on Mac.

Tags

The Notes app in iPadOS/ iOS15 gives individuals the ability to create tags to conveniently sort and organize notes in a new approach. We have a Tag Browser to click combinations of tabs and easily view tagged notes and new custom folders that automatically house notes depending on tags.

Reminders

This app now comes with tag features that enable you to organize the reminders, and you can find and filter reminders depending on tags. We have a Tag Browser to click combinations of tabs and easily view tagged reminders and new custom folders that automatically house them depending on tags. With iPadOS/ iOS 15, we have quick-access options to seamlessly delete completed reminders, enhance natural language support, and expand suggested characteristics such as flags, priority, tags and much more.

FaceTime Links

In iOS/iPadOS 15, you now have the ability to create a link to a FaceTime call and share it through Mail, Calendar, Messages, and third-party apps like Facebook, WhatsApp, etc. You can open FaceTime links to access the FaceTime app in your Apple devices, however these links can also be launched via a web browser. As a result, introducing FaceTime to Windows and Android for the first time. FaceTime calls on the web remain end-to-end encrypted for security reasons.

Messages

Under the new option Shared with You of a corresponding app sent Messages automatically shows up there. Apps such as Apple News, Photos, Apple Music, Podcasts, Apple TV, and Safari all support the new feature Shared with You. Also you can pin a prestigious item that has been sent to you, so that it is placed above in Messages search, Shared with You, and the Details view of the conversation. Clusters of images sent via Messages will be displayed in a collage at a glance or in a stack, which can be swiped based on the numbers of images sent. And it is easy to save images with the new save toggle. With a contact name you can easily access pictures shared via Messages.

Photos

In iPadOS/iOS 15 there are improvements to the Memories tab of the Photos app. The tab is now integrated with Apple Music, comes with a new design, an additional interactive interface, and Memory looks.

Depending on your Apple Music listening history that synchronizes with images and videos for additional customized experience the Memories tab will recommend songs. You can personalize the Memories tab by swiping through Memory mixes, helping you to try out separate songs with a distinct tone and momentum.

With iOS/iPadOS 15, we have up to 12 Memory looks that introduces ambience through the analysis of every video and picture, and administering the right quantity of color and contrast adjustment for a steady appearance.

Apple also designed new Memory types, which include, child-focused memories, improved pet memories, additional international holidays, and trends over time, coupled with the capacity to identify different cats and dogs. You can also see and adjust all of the items from a Memory in a bird's-eye view, and a Watch Next window to put forward interconnected memories to view.

Also with these updates, the feature **People identification** presents improved recognition for users. As a result it is straightforward to correct names wrongly spelt in the People album.

In iPadOS/iOS 15 Photos' app you also have the option **Feature Less** that informs the app that you want to view less of a certain place, date, holiday, or person across Featured Photos, the Photos widget, Memories, and selected in the Library tab.

The Photos app now offers a more detailed Info window to access the property of the image, like the lens, size, camera, and shutter speed, or sender of Shared with You photo in Messages. It is also possible to edit the date taken or location, add a caption, and learn about content recognized by Visual Look Up.

Also the Photos app image picker is integrated with the Messages app which lets users choose pictures in a particular order for sharing. If you grant third-party apps access to some items in the library they can in turn deliver easier selection workflows

And according to Apple the initial sync of iCloud Photos on a new device is faster on devices running iOS or iPadOS 15.

Maps

In iPadOS/iOS 15, the Maps app features a new 3D city-driving experience with road facts like bike lanes, medians, turn lanes, and pedestrian crosswalks, which functions both on the iPhone and Apple CarPlay.

Maps also offers interactive globe view with exceptional improved details in a new 3D view for cities.

New road names and colors, custom-designed landmarks, and a new "moonlit" night-time mode alongside elevation, commercial districts, neighborhoods, buildings, and much more are now shown in detail.

Transit navigation in Maps has been restructured and users can search for nearby stations effortlessly and pin their favorite lines. Maps will automatically follow along a highlighted transit route, prompting you when it is almost time to land.

Based on the restructured place cards it is very easy to spot and communicate with information about locations, physical features and businesses. We also have a new Guides Home that gives editorially and carefully selected details of the new area.

When looking for a new site, there are new options to filter results by criteria like cuisine or opening times. Maps will also automatically update search results when moving around and the most-used settings are now located in one, easy location.

App Store

With iPadOS/iOS 15, the App Store now offers an easy channel to locate current events across games and apps coupled with movie premieres, live streamed, or game competitions experiences. A new App Store widget displays collections, stories and in-app events from the Today tab.

Screen Time

In iPadOS/iOS 15, you have can to switch to downtime when required, only the apps and phone calls you decide to enable will bypass the system. Once turned on, a 5-minute downtime reminder will be sent and it will remain active all through the whole day. Developers can make use of Screen Time API in parental controls apps to support various parental tools. The API offers developers with the features such as core restrictions and device activity monitoring.

Privacy

In iPadOS/iOS 15, Apple took privacy to the next level with new transparency features, controls, and protections. The Mail Privacy Protection enables you to stop trackers from spying on your email as well as stop them from accessing your IP address. And App Privacy Report throws light into how apps are sharing information with other companies. Additional new feature in Settings is the App Privacy Report, which allows users to see the way frequently used apps access their camera, location, contacts, microphone, and photos, within the last 7- days. Via this feature you will know apps that have interacted with your domains and when last they contacted them. In these updates Developers can enable users to paste items from a different app without accessing copied items until you want to. And Developers have the ability to let users share their current location momentarily alongside an adjustable button in their apps. Now Developers can give smart functionality when accessing the Photos library, asking for limited access to certain photos folders and album selection.

iCloud+

iCloud+ integrates everything you see in iCloud with new premium features, including Hide My Email, expanded HomeKit Secure Video support, and an innovative new internet privacy service, iCloud Private Relay, at no extra cost. For iCloud subscribers, you can easily upgrade to iCloud+ automatically this fall. All iCloud+ plans can be shared with individuals with one same Family Sharing set up, so that every member can explore the new features, storage, and elevated experience that comes with the service. The iCloud Private Relay feature lets you browse with Safari while encrypting all traffic from your device. Requests are conveyed via two distinct internet relays and it is so, to prevent anyone from accessing your IP address, location, and internet activity to build up a comprehensive profile of you. With Hide My Email you can design uncommon and random email addresses that redirect to your personal inbox, so that you can send and receive emails without having to share your real email address. With iPadOS/iOS15 you can now personalize your iCloud Mail address with a custom domain name, and then invite family members to make use of the same domain with their iCloud Mail accounts. With your iCloud+ account you can save video clips from multiple security cameras and do not contribute toward your iCloud storage allowance.

Apple ID

With iPadOS and iOS 15 you can now choose one or more trusted individuals to become an Account Recovery Contact to help reset your password as well as account recovery.

The new Digital Legacy program lets users pick individuals as Legacy Contacts, who can access your account and personal data in case of accident or death.

Accessibility

With iPadOS/ iOS 15 you can now examine objects, text, persons and tables within photos thoroughly through VoiceOver.

And with the help of the VoiceOver image descriptions in Markup you can include image descriptions that can be read by VoiceOver. Image descriptions continue even if shared and can be read in several compatible apps on Mac, iPad and iPhone.

The New accessibility features in iOS/ iPadOS 15 are configured to function the way you work. The VoiceOver screen reader now uses built-in intelligence to analyse objects within pictures, allowing you to find out comprehensive information about an individual, text, table data, and other objects within the images. The Support for third-party eye-tracking hardware helps people to manage iPad through their eyes. Background sounds run steadily and mix into or duck under other audio and system sounds to hide unwelcome surrounding noise as the device is working. Thereby helping you stay calm, restful or focused. Sound Actions enable you to personalize Switch Control to function with mouth sounds. You can now personalize display and text size on an app-by-app basis.

With these updates, the Magnifier app is now a default app, and an iPad can now be used as a magnifying glass to zoom in on objects close to you.

We now have new Voice Control languages as Voice Control has brought in new language options such as Cantonese, Mandarin Chinese, German, and French, that use Siri speech recognition technology.

Wallet

In iOS 15, the Wallet app supports remote keyless entry control, thereby letting users honk their horn, unlock or lock their car, open the trunk or preheat the car. The app widened its compatibility for vehicle keys, via Ultra Wideband chip it can unlock, lock, and switch on the ignition without having to pull out your device from your pocket. The Ultra Wideband chip offers accurate spatial awareness, meaning that the operating system will restrain you from locking the vehicle while your device is inside or start your vehicle when your phone is outside. Beginning later in the year 2021, users living in partaking states in the United States can add their state IDs or driver's license to the Wallet app. The Transportation Security Administration is up and running to establish airport security checkpoints to be the first platform users can make use of the digital Identity Card in Wallet. With these updates this app can now automatically register outdated boarding passes and ceremony tickets.

Safari now supports adding numerous passes to Wallet in one entry instead of manually adding them individually. Also in these latest updates the Wallet supports other kinds of keys, like hotel room key cards, office, or home.

Find My

Find My comes with the potential to assist in discovering In iOS/ iPadOS 15, Find My comes with the potential to assist in discovering lost devices with the aid of the Find My network even if the device has been switched off or erased. Users now have the ability to steadily live-stream the location of family members or friends who desire to give out their location with them to offer a sight of guidance and speed. The Find My network now supports AirPods Max, and AirPods Pro. We have a new Find My widget for at-a-glance look at places. Also you will get a new Separation Alerts to prompt you when you forget an Apple device, AirTag, or Find My accessory network in a strange place. Siri In iOS/iPadOS 15, Siri requests are carried out via the help of an on-device Neural Engine, increasing security and impressively enhancing sensitivity. It can now function without internet connection. Built –in understanding and speech recognition gets better 1as you engage the tablet. Siri will comprehend contacts you frequently communicate with, new words you enter, and subjects you study about to deliver better responses. You can now share on screen content such as web pages, images, items from Apple Podcast, Apple Music, Apple News stories, Maps locations, and much more via Siri in a Message. Also, via Siri you can employ on screen context to initiate a call or send a message. So you can now conversationally refer to previous requests based on Siri being stronger at sustaining context between requests. Users can also create requests to direct HomeKit accessories at fixed time or within defined conditions, such as when you are on vacation. Siri can now proclaim Notifications, like Reminders, on AirPods and Siri can be asked the content of your display.   With iOS/ iPadOS 15, Siri now features neural text-to-speech voice in numerous languages including Norwegian, Finnish, Danish, and Swedish. Siri also has language support for Mixed English, Indic, a mix of Indian English and a native language.

5G Connectivity

The iOS/ iPadOS 15 comes with much improved connectivity via 5G. Several apps and system experiences are refined through high-speed 5G on supported devices, including the support to back up to iCloud and restore from an iCloud backup, stream audio and video on Apple and third-party apps, download higher-quality Apple TV+ content, sync photos to iCloud Photos, and many more. With iPadOSiOS 15 5G is favoured over Wi-Fi – Apple tablets like the 12.9-inch iPad Pro (5th generation) and 11-inch iPad Pro (3rd generation) will now automatically line up 5G when the Wi-Fi networks you tethered to is often slow, or insecure.

Memoji

The iOS/iPadOS 15 added about nine new Memoji stickers to enable users to send a light bulb moment, a hand wave, a shaka, etc. And we now have more than 40 new outfit choices with up to three color combinations, including for headwear. In the latest iPadOS/iOS Memoji supports a different color for your right and left eyes. There are now three new glasses options which include retro shapes, star and heart. Also included in these updates is representation of soft helmets, oxygen tubes, and Cochlear implants in the Memoji menu.

Gaming

With iPadOS/ iOS 15, your newest Messages from friends and group are conveyed into the Game Center-enabled games alongside a new multiplayer friend selector. The Game Center can now display imminent demand in the Game Center friend request inbox. In these updates you now have the potential to save a video footage of up to the last 15 seconds of gameplay by clicking the share button on game controllers such as the Xbox Series X or Series S Wireless Controller, or Sony PS5 Dual Sense Wireless Controller. We have a new Continue Playing widget that shows latterly played Game Center-enabled games across iPad. A Friends Are Playing widget enables you to know the games their friends play. A recommended Focus for gaming allows individual stay engrossed in games by removing unwelcome notifications.

Music

With the new Spatial audio alongside dynamic head tracking feature in iPadOS/iOS 15 users can employ AirPods Max and AirPod Pro to listen to songs with Dolby Atmos and Apple's dynamic head tracking for a more engulfed experience. And with this latest Music app user can decide to select music shared with you from Messages.

Podcasts

In iPadOS and iOS 15, Podcasts develop customized groups of recommended shows about specific topics.

You can now share your favorite podcast episodes in the Messages app and get every episode shared with you in the **Listen Now** tab.

News

In iPadOS and iOS 15 the News feed has been redesigned which make it easy to navigate and communicate with articles. Information like bylines and publication dates are notable and stories can be saved and shared right way from the feed.

Stories received on the Messages app will now automatically pop up in the **Shared with You** interface in the Today and Following Apple News tabs.

TV

Apple introduced a new-fashioned row called For All of You in iOS/iPadOS 15 and it is built to put forward a number of movies and TV shows derived from the passion of a whole family or certain individual. The TV app also features highlights of every movie and TV show that family and friends have shared in Messages app. Also, with iOS/iPadOS 15 this app works seamlessly with Messages and FaceTime apps to view content with family and loved ones in sync via SharePlay. For those in Asia nation of Japan, you now access common streaming apps via the Apple TV.

Voice Memos

You can now slow down or speed up playback of recordings done via the Voice Memos app in iOS/iPadOS 15. The app automatically examines recordings and automatically skips over gaps in your audio with a single tap. You can now share multiple Voice Memos recordings at once.

Shortcuts

With the Next Action Suggestions feature in iOS/iPadOS 15 you can build shortcuts on your device. And Shortcuts can initialize with iPhone, iPad, and Mac. Via Cross-device management option users can share Shortcuts and download alongside a link, without having to control the security settings. Recipients will now get notification to make sure that only permitted information is shared.

Weather

Apple overhauled the Weather app in iOS/iPadOS 15 and it now delivers fresh graphical displays for weather data, a full-screen map, and a dynamic layout that switches depending on external surroundings. Also the Weather apps dynamic backgrounds perfectly mirror the sun's current position and precipitation conditions. Users will also get alerts that will announce when rain or snow will start and cease. Also the Weather app's animated backgrounds were restructured to perfectly show the sun's current position and precipitation conditions. You also have notifications to select when snow or rain starts and come to an end.

Health

In iOS 15, this app comes with a new sharing tab that enables users to share some health data with medical personnel or loved ones. Lab results also got improvements with descriptions, highlights, and the option to pin results for quick access. The Health app can now identify Trends, drawing observations to meaningful changes in personal health metrics. Also it comes with Walking Steadiness as a new metric to help shape fall risk. COVID-19 immunizations and test outcomes can be saved in the Health app through a QR code from a healthcare provider. Blood glucose highlights now display levels during sleep and exercise, and show interactive charts.

Camera

Improved Panorama captures - Panorama mode in iOS 15 now comes with enhanced geometric distortion and can offer beautiful shots of moving subjects while also minimizing banding and image noise.

In iOS 15 you can swipe down or up while taking a QuickTake video to zoom out or in.

Apple Card and Apple Pay

Apple Card customers can have a security code that switches habitually to secure online Card Number transactions.

Also, you can easily and swiftly access your card number by launching your Apple Card in Wallet and clicking the card icon.

A re-engineered Apple Pay payment sheet lets you add new cards and coupon codes inline. Also we have an improved summary view that displays elaborate information, such as discounts, subtotals and payment items.

Communication Safety

In iOS/iPadOS 15, we have a new Communication Safety functionality in the Messages app. It is designed to notify children and their parents when sharing or receiving sexually crystal-clear images. The Messages app will utilize on-device machine learning to scrutinize photo attachments, and if an image is ascertained to be sexually obvious, the picture will be automatically blurred and the child will be cautioned.

If a child makes an effort to access an image that has been tagged as sensitive in the Messages app, they will be prompted that the image might have private body parts, and that it is destructive. Based on how old the child is, there is also an option to notify parents when the child goes ahead to access the sensitive picture or if they decide to share a sexually sensitive picture with an individual after being admonished.

Devices Compatible with iPadOS & iOS 15

These updates can run in the same iPhones as iOS 13 and iOS 14, including older devices such as the initial iPhone SE and iPhone 6s, iPhone X, iPhone 8, iPhone 8 Plus, iPhone 7, iPhone 7 Plus, iPhone 6s, iPhone 6s Plus, iPhone SE (2016) and iPod touch (7th generation). And on the other hand the iPadOS 15 is can run in all of the same iPads as iPadOS 13 & 14, comprising older devices such as the first-generation iPad Pro, iPad mini 4, and fifth-generation entry-level iPad. The impressive thing about iOS/iPadOS 15 is the seven years of software support, that is all iPads from 2014 are compatible with iPadOS 15.

Installing iPadOS/iOS 15 in Your Device

To install your iPadOS or iOS 15 make sure you backup the iPad or iPhone. And you have several ways to back up your device and you can do this either in iTunes, Windows, Finder on a Mac or iCloud. And if you have enough iCloud storage, it is the easiest channel to back up your device.

To do this;

- Go to the **Settings app** on your device.
- Click on your name or profile icon at the top of the Settings interface.
- Select **iCloud**.
- And under **iCloud** you will see numerous settings and apps using iCloud. Then scroll down and tap on the option **iCloud Backup.**
- Then tap on the link **Back Up Now**.
- Alternatively, you can back your device via a lightning cable plug into a computer.
- And once you are satisfied that you have a backup that is up to update, launch **Safari**.
- Go to Apple website **beta.apple.com** or **apple.com/beta.**
- You see all the beta available and if you have not signed up with your Apple ID go ahead and do that. For example if you have already signed up from previous tap the link sign in.

- Once you are in, highlight the beta you want, iOS, iPadOS, macOS, tvOS or watchOS.
- Scroll down to where it says **enroll your device**. It will take you through a few things on how to back up your device. And click the link, **enroll your iOS or iPad device** based on the device you selected earlier.
- The system will once again remind you of the need to have a backup. And tap the link **Download profile.** By tapping this link Apple is prompted that you need the beta of the OS on the device.
- Wait for a few seconds and you will get a small dialogue box and then tap **Allow**.
- When a small screen with the saying **Profile Downloaded** pop up click on **Close**.
- Return to the Settings screen and just beneath your profile icon or name you will see **Profile Downloaded.**
- Tap on the link **Profile Downloaded.**
- Click on **Install** at the top right of the Install Profile display.
- Enter your **passcode**.
- At the top right of the Consent screen tap on **Install** again and go to the bottom to toggle **Install** once more.

- You will be prompted to restart your device via a small screen. Go ahead and tap **Restart**. And this will install the beta profile.
- When your device is back On, go to **Settings** and select **General**.
- Tap on Software Update. It will take a moment and ensure you are on Wi-Fi because the download is slightly huge. However you might be able to use cellular depending on your location.
- Give it a few seconds for the device to check. And ensure the **Automatic Updates** option is set to **Download Only.** And if you enable **Automatic Updates** you will not find the updates, so it is better you disable it.
- Tap on **Download and Install.**
- And once your device installs the iOS or iPadOS 15, reboots to explore the OS.

Have this in mind that this is a beta version, so it might have some issues. And to revert back to iOS/iPadOS 14, you will require a computer to carry out the task. You will need a Mac using Finder or a Windows computer using iTunes.

Downgrading to iOS 14 from iOS 15

Using the beta versions of iOS/iPadOS are always characterized by bugs and numerous issues especially the first release. So if you've switched to the recent updates via the public beta or developer program of Apple. Apps might not be running properly, device crashes, poor battery life, and features not performing their purposes. The good thing is that it can be restored to iOS or iPadOS 14 as a result of the buggy nature of the updates. The backup you created before installing the iPadOS or iOS 15 beta can be restored when restoring back to the iOS or iPadOS 14. Understand that you can return to iOS or iPadOS 14 without having to create a backup, however you cannot restore the device to the point it was when you first installed iPadOS or iOS 15 beta. And if your Apple Watch is running watchOS 8, it will not work with your iPhone when you downgrade to iOS 14.

To downgrade your device operating system;

- Plug the device to a Mac via a Lightning cable.
- Open the **Finder** application on the Mac.
- Switch the device to **Recovery** mode. The process of launching the recovery mode varies. For those with iPad with Home button, iPhone 6s or earlier, and iPod touch (6th generation) or earlier: Long-press the Home and the Side/Power buttons together. Don't release the buttons until the recovery mode display shows up. For users with iPad models with Face ID, iPhone 8 or

later: tap and quickly release the Volume Up button. Tap and quickly remove your finger from the Volume Down button. Long-press the Power button until your device restarts. Hold down the Power button until the device switches to recovery mode. And for persons using iPhone 7, iPhone 7 Plus, and iPod touch (7th generation): Press and hold the Side/Power and Volume Down buttons at the same time. Don't release buttons until the recovery mode window pops up.

- On the small screen that shows up, tap **Restore** to wipe the device and install the updates public release of iOS or iPadOS. It will take few minutes for the process to be successful.

SECTION TWO

How to Set Up Your Device

Apple introduces so many features in iOS / iPadOS 15 and if you don't know where to start from, it is ideal you set up your device first. From here you can explore all the tons of tricks, tips, hidden features and news functionality in iOS / iPadOS 15.

When you turn on your device you will get the Home welcome page with a handwriting like Hello, Hola, Helo, Salut, or in other multiple different languages.

To complete the set up from a device that is already activate;

- Press the Home button or swipe up on the home screen.
- Enter your passcode.
- The Software Update Complete screen will show up. And then tap **Continue** at the bottom of the screen to proceed.
- On the **Terms and Conditions** display tap **Agree** at the top right corner. Tap Agree again on the small dialogue box that will pop up.
- On the Apple Pay screen click on **Continue** or you can tap on **Set Up Later in Wallet.**
- Tap **Continue** on the iPhone/iPad Analytics screen.

On the Welcome iPad or iPhone page tap on **Get Started.** And that will take you to the Home Screen and you can now begin exploring the iOS/iPadOS 15 updates.

However, for those who are setting up their device as a new one;

- Press the **Home button or swipe up** on the slide bar at the bottom of the display. And select your language, **English** for example.
- Choose your region or country. **United States** or **India** for example.
- The next screen is for **Quick Start**. The device will prompt you to bring your current iPad or iPad close to the new device. Once the older device is close to the new one, you will see **Set Up New iPad/iPhone** on the older device screen. If no other device, tap **Set Up Manually**.
- On the Written and Spoken Languages screen you have the option to set up Preferred Languages, Keyboard and Dictation via the option Customize Settings. Or press **Continue** to proceed.
- Select a Wi-Fi network. And it will take some time to activate the device.
- On the Data & Privacy page tap on **Continue**.
- Next line of action is to set up the Face ID or Touch ID depending on the device. You can **tap Set Up Face ID /Touch ID Later** to skip face ID/touch ID set up. And to proceed, toggle **Continue**.

- On the Apps & Data screen you will see choices of how you can transfer apps and data to your new iPad. You have the option **to Restore from iCloud backup, Restore from Mac or PC, Transfer Directly from iPad, Move Data from Android, or Don't Transfer Apps & Data.** Tap on **Don't Transfer Apps & Data** to set up the device as a
- Enter your **Apple ID** details.
- Tap **Continue** on the Apple ID Security page.
- Tap **Agree** on the bottom right corner of the Terms and Conditions screen. You then see on the screen that it will take some minutes to set up your Apple ID.
- Click **Continue** on the Keep Your iPad/iPhone Up to Date screen.
- Enable Location Services.
- On the Screen Time display toggle **Continue** to move forward.
- On the iPhone/iPad Analytics screen tap **Continue**.
- On the Appearance page, select **Light** or **Night** mode and then hit **Continue** to proceed.

On the Welcome iPad or iPhone page tap on **Get Started**. And that will take you to the Home Screen.

Home Screen Customization

The first thing you will observe on the Home screen is that the Today view is no longer there as we have it on the left hand side in iPadOS 14. And if you swipe to the right from the left side of the screen you will see some widgets. However you now have widgets on the Home Screen and it is a new feature in iPadOS but not new in iOS. To bring in widgets to the Home screen to stay together with apps on the Home screen:

- Long-press on the blank space on Home Screen.
- Click the Plus sign (+) on the top left corner. Then you notice that we have a brand new widgets UI. And on the left of the Widgets interface we have a sidebar with the lists of widgets we can select from. In addition we have five new widgets which include the App Store, Find My, Game Center, Mail and Contacts.
- Once you tap a particular widget on the left side bar, swipe over on the widget at the middle of the widgets interface to see different styles.
- And long-press on the select widget and drag it to the Home screen. Alternatively you can tap the option Add Widget.

And you can reorder widgets to anywhere on the Home screen by dragging them.

WPA3 Security

The iPadOS/iOS 15 Hotspot Connections offers a secure WPA3 security protocol. This comes with a well enhanced security and updated procedure in place to frustrate password predicting. WPA3 was first launched by the Wi-Fi Alliance in June 2018 with the aim to simplify Wi-Fi security, offering tough authentication, and improved cryptographic efficacy. In iPadOS/iOS 14, hotspot connections from an iPad or iPhone are only guarded by WPA2. And now with iPadOS/iOS 15 hotspot connections will be protected additionally with stronger WPA3. For some time now, Apple's devices can pair with networks that have WPA3 security but up to this point, personal hotspot networks established by those devices are only compatible with the ancient and feeble WPA2 standard. The hotspot connection know-how on iPadOS/iOS 15 might remain consistent for normal users. Regardless of advocacy to design vigorous, advanced and complex passwords for hotspot networks, most people will keep working with simple passwords which could easily be guessed. The WPA3 security protocol is one that is password-based, extra resilient and authentication that renders powerful protections for people in opposition to password guessing trials by a different person somewhere or hacker.

Resetting Button

With iOS/iPadOS 15 the Rest Button is now **Transfer or Reset iPhone/iPad**. And to locate the button, go to the Settings app and under **General** you will see the button. All of the reset options for Keyboard Dictionary, Home Screen Layout, Network Settings, etc, are now placed under the **Reset** button.

Using Background Sounds

An exciting accessibility feature in iOS/iPadOS 15 is Background Sounds, and it is formed to enable users to remain calm, focused, and lessen interference with assistance from your device. Also the Background Sounds delivers bright, dark noise as well as natural sounds such as stream, ocean and rain. Every of these sounds can be enabled in the background to block unwelcome surrounding or outside noise, and the sounds immersed into other audio system sounds. To activate Background Sounds;

- Go to the **Settings** app on your device.
- Select the option **Accessibility**.
- Under the heading **HEARING**, click **Audio/Visual.**
- Toggle **Background** Sounds.
- Click the switch next Background Sounds to turn on/off. Hit the option Sound to select the sound effect you want. And in the Sound window you

can select Balanced Noise, Bright Noise, Dark Noise, Ocean, Rain, or Stream.

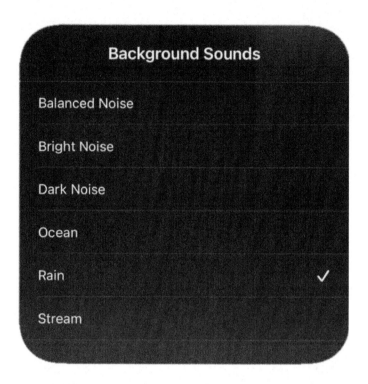

Also for quick access you can bring in the Hearing shortcut to Control Center via the Control center interface in the Settings app and from it can easily be opened. You have to download these sound effects separately when you play them for the first time. Ensure are connected to internet and you can play it offline after downloading.

Shared with You

The major enhancement Apple brought to the Messages app in iOS/iPadOS 15 is the feature called **Shared with You**. It helps to accumulate items you received from other people via the Messages app into the conventional apps. For example, when you receive an image from an individual it will pop up in the **Share with You** tab of the Photos app. This feature is not only designed for the Messages app, however it makes Messages items easily obtainable.

Other apps that are incorporated **Shared with You** are; Safari, Photos, Apple TV, Apple News, Podcast, and Apple Music.

For example if someone shares with you a TV show or film link via the Messages, the videos file will show up in the Share with You tab of Watch Now in theTV app for easy access.

When you receive a link in Safari, the link is saved in the **Shared with You** area of the primary start page which is launched when a new tab is opened. Via Safari you will see a preview of the link and you click the link to go to the website.

Another example is that, the Music links sent in Messages are displayed in the **Shared with You** segment of the Apple music in the Listen Now tab.

How Focus Mode Works

The Focus mode is the modification of the Do Not Disturb feature and it was designed to lessen distractions. The new Focus mode in iPadOS /iOS 15 is compatible with Mac and Apple Watch. It introduces flexibility, control, automation into your device and filters messages, set notification, call, and auto-reply to others in your absence.

With these updates running on your device you can focus on some specific features based on your preference.

You can focus on Do Not Disturb if you want to wind down in the night as well as fitness, work, study or other numerous customization.

In a nutshell, Focus mode enables you to decrease distractions thereby allowing you to focus on a specific task.

How to Enable/Disable Focus Mode.

You can turn on a Focus mode on your device in order to minimize disturbances to pay full attention to a particular engagement.

To do this;

- Pull down the **Control Center** by swiping down from the top-right corner of the device's screen.

- Click the **Focus button** and then choose the Focus you want to enable by selecting it.

- Tap the ellipsis (…) at the top-right corner of the selected mode window to enable it. **For 1 hour**, **Until this evening**, or **Until I leave this location**.

And to disable Focus Mode, return to the Control Center and click on the Focus button. Then toggle on the highlighted/active focus button to turn it off.

Note that when you disable/enable a Focus mode, it will automatically sync with all your Apple devices. So to change this general thing, go to the **Settings** app, select **Focus** and disable the option **Share Across Devices** by clicking the switch next to it.

Turning On Smart Activation for Focus Modes

The Smart Activation option in Focus modes allows you to automatically activate individual Focus modes at favorable times throughout the day depending on indicators such as app usage, location, etc.

For instance, when you have set up a custom Focus mode known as **Fitness** and you normally do workouts with the Fitness app, or in the gym. The Smart Activation will master your routine. And the device will knowingly trigger the created Focus mode without you having to take any action.

You can set up Smart Activation Focus mode individually and to do this;

- Open the **Settings** app.
- And click the option **Focus**.

- Tap the **Focus** mode you desire to automate.
- Beneath the heading **TURN ON AUTOMATICALLY,** select Smart Activation.
- Click the switch next to Smart Activation to turn it on/off.

How to Schedule and Automate Focus Modes

The new Focus feature in iOS/iPadOS 15 allows you to concentrate on a certain work at a time. Focus mode does this by filtering notifications depending on your task and you can customize it for certain conditions.

Apart from turning on Focus mode on your device to minimize interference and concentrate on specific jobs. You can also schedule and automate Focus modes to be triggered at the set time of the day, working hours, when in specific locations like the office or church, when winding down to sleep or when you launch certain apps on your device.

- To schedule and automate Focus mode;
- Open the **Settings** app on your device.
- And then select **Focus**.
- Choose the **Focus** mode that you want to schedule.

- Beneath the heading **Turn On Automatically**, click **Add Schedule or Automation.**
- Set Time, Location, or App, based on what you want. Under **Time** you have to set the duration via the **From** and **To** options. Select days of the week you want the mode to be activated via the option **REPEAT**. For App, select the apps that you want the Focus to affect and click **Done** to confirm. And as for **Location,** browse and key in the address via the entry field and toggle **Done** to confirm

Creating a Custom Focus

The Focus Mode comes with several pre-installed modes such as Personal, Work and Sleep. However, with iPadOS /iOS 15 you can create multiple focus modes and generally you can set up what happens when you are doing a particular thing. This will silent your notifications and you can choose apps and people you want to receive notification from. You can also schedule time so that the feature turns on automatically when it gets to the set time of the day.
To your Custom Focus mode;
- Go to the Control Center and they click the Focus button.
- Toggle the **New Focus** button at the bottom of the Focus mode screen.
- Under the heading **Choose a Focus to get started**, click **Custom** to create a new Focus.

60

- Name the custom Focus and pick an emoji, icon or color for recognizition purpose.
- Click **Next** to proceed.
- Click the link **Add Person** to select the individual you want to get alerts from while Focus mode is activated.
- And via the link **Calls From** you can decide to receive calls from Favorites, No One, All Contacts, or Everyone.
- Select the option **Allow [number of person] Person** or **Allow None** to proceed.
- Click **Add App** to select the apps you want to get alerts from when the Focus mode is turned on.
- Select the option **Allow [number of apps] Apps** or **Allow None** to proceed.
- Next thing is to decide whether to **Allow Time Sensitive notifications** when the Focus is ON or click **Not Now** to set it in future.
- Toggle **Done** to finish creating the custom Focus.

You can adjust the Focus mode you created via the **Settings** menu under the option **Focus**. Access the custom Focus by its name. You can decide to customize the options like Hide Notification Badges on app icons, Dim Lock Screen appearance and Show On Lock Screen any silenced alerts you get. Also you can choose to hide a specific Home Screen pages via the Custom Page switch.

- To delete a custom Focus mode;
- Go to the **Settings** app on your device.
- Click **Focus**.
- Choose the Focus you want to delete via the name.
- Scroll to the bottom of the screen and hit **Delete Focus**.

Enabling Time Sensitive Notifications

The Focus mode in iPadOS/iOS 15 helps you to stay focused while doing a particular work by reducing disturbances and stopping prompts that are not connected to the work.

But in a situation where you are expecting very vital information from an organization, friends, boss or family. While the Focus mode is still very much active, key notifications can still enter your device via the option **time sensitive notifications.**

Notifications designated as time sensitive can get around or bypass Focus mode.

Also you can decide which modes get time sensitive notifications and to enable the feature;

- Go to the **Settings** app on your device.
- And then click the option **Focus**.

- Highlight the Focus mode you desire the time-sensitive notifications to bypass.
- Tap the switch next to Time-Sensitive Notifications to turn if off/off. Note the green indicates ON.

Using Quick Notes

Quick Notes is a brand new productivity feature in iPadOS 15 that offers smart means to put down things on your device without having access to the Notes app. For those using a device plugged to a keyboard with a Globe key, just tap the Globe key + Q to fire up Quick Note.

Alternatively, the Quick Note button can be added into the Control Center. To do this; open the **Settings** > **Control Center** and under the **Included Controls** area add the Quick Note. However, previously used notes come up when you trigger Quick Note, so continue adding notes to display a fresh one. And you can create a fresh one by clicking on the New Note icon at the upper right corner of the Quick Notes display.

Regardless of the page you're on, whether in an app or on the Home screen, a floating Quick Note interface can be initiated whenever you want, **by** swiping diagonally up from the bottom-right corner of the display with an Apple Pencil or finger .

Normally, a virtual keyboard designed for typing will pop up when you click on a note. For those with Apple Pencil , you can turn on the **Scribble** feature via the **Apple Pencil** settings in the Settings app. So with your Apple Pencil you can put down handwritten notes and they'll be transcribed into conventional text.

At the upper right corner of the Quick Note page you will see the quadrant icon ⠿ that gives you access to the main Notes app menu, where you have all your Quick Notes. Next to the Quadrant icon we have the **Ellipsis** ● icon, designed to share notes and the new note icon.

Also you will see a Markup icon ⬤ at the lower-right corner on the Quick Notes display. This icon will bring out the habitual Markup tools at the bottom of the screen. Usually you have an Ellipsis icon at the around lower section of the currently considered note when you have formed numerous notes. This Ellipsis icon helps you to swipe between the notes you created inside the editor interface.

Recently used apps or the ones you're currently making use of can be detected on Quick Notes via a dropdown menu at the top of the editor interface. This menu is designed to add links to certain messages you're viewing in an app and whenever you're viewing a note with a link, simply click the link and you'll be led directly to the related content.

You can drag the Note screen to any corner of your device's display when it is blocking your view.

If you want to set aside the Quick Note swipe diagonally towards the corner of the screen it's nearer to and you can bring it back whenever you want by pressing the let out bar off to the side. Alternatively, click on **Done** at upper- left of the Quick Note screen to put away the feature.

Be informed that you cannot fire up Quick Notes on iPhones powered by iOS 15, but you can get your notes created in iPad or Mac on your iPhone since Quick Notes dwell in the Notes app.

Using Siri Offline

The entire speech processing and personalization functionalities of Siri has been moved onto your device, thereby aiding Siri to be fast in commands processing and becoming a robust voice assistant. What this implies is that the virtual assistants can now run several commands completely offline. If iOS/iPadOS 15 is running on your device, then you don't have to toggle any switch for the virtual assistant Siri to function offline. Some of the operations Siri carry out offline includes; opening apps, setting up alarms and timers, turn on/off alarms and timers, Control system settings such as volume, Low Power mode, accessibility features, Airplane mode, etc. managing Podcasts audio playback and Apple Music. Whenever you ask Siri to process a task that requires internet connection and if you're not connected to a

Wi-Fi network connection or cellular data. Siri will tell you that I can help with that when you're connected to the internet or to do that, you'll need to be online.

Communication Safety

In iOS 15/iPadOS 15, Apple introduced a new-fashioned child safety measure and it is designed to make the internet safe for children. Though some individuals are not comfortable with the feature because of privacy anxiety as it scans photos shared and received via the Messages app for sexually crystal-clear content. However, Apple emphasized that this is a participative functionality confined to the accounts of toddlers. It should be turned on by the child's guardian or parents via the Family Sharing menu. When Communication Safety is enabled for the Apple ID account of a kid, Apple will inspect the photos that are shared and received in the Messages app for sexually harmful material. When sexually harmful material is found, the image will be automatically blurred and the toddler will be cautioned that the picture might have naked content. You will see sexually harmful content. It's not your fault, but sensitive photos and videos can be harmful to you. However the child can decide to view the image. And for a child who is under the age of 13 their parents can decide to receive a notification when their child bypasses Apple's warning. "If you resolve to access this, your parents will be notified to ensure you're OK," will be displayed on the screen. The parental notification functionality is optional and is only present when the kid attempting to access the image is below the age of 13. When children within the age 13 - 17 are doing the same thing their parents will not be notified, but they will receive the warning about sensitive content when Communication Safety is enabled. The Communication Safety feature is also accessible for persons below the age of 18 and cannot be activated on adult accounts. Parents will need to opt in to Communication Safety when setting up their devices via Family Sharing, and it can be turned off. The Communication Safety feature makes use of on-device machine learning to process image attachments.

Child Sexual Abuse Material (CSAM)

Apple are really pushing the safety of children to next level and with iOS/iPadOS 15 your device can detect Child Sexual Abuse Material in the United States.
With this new feature Apple will be able discern notable Child Sexual Abuse Material pictures saved in iCloud Photos, thereby aiding Apple to communicate these occurrence to the National Center for Missing and Exploited Children (NCMEC). The CSAM is configured with user privacy in mind and Instead of checking out pictures in the cloud, the setup will conduct on-device harmonizing against a database of

recognized Child Sexual Abuse Material pictures hashes delivered by the National Center for Missing and Exploited Children and other child safety bodies. In addition, the system will metamorphose this database into an indecipherable collection of hashes which are encrypted with an iPhone or iPad. The 2Child Sexual Abuse Material photo scrutinizing occur automatically and it is not an optional function. However, Apple buttressed that it will be unable to dig out Child Sexual Abuse Material pictures when the Photos functionality is disabled. The blueprint of the whole idea is that, the system recognize a well-defined Child Sexual Abuse Material photo on a device and then point it out when it's transferred to the Photos with a token assigned. After a specific amount of tokens have been sent to the Photos, Apple can then unravel the tokens followed by manual examination. When Child Sexual Abuse Material item is identified, the user account is deactivated and a signal is afterwards conveyed to the National Center for Missing and Exploited Children. NeuralHash, a hashing technology, studies a photo and turns it to a special number exclusively for that photo. The primary aim of the hash is to make sure that homogeneous and visually alike pictures result in the same hash, while photos that are distinct from one another result in different hashes. For instance, a photo that has been edited, resized, cropped or converted from color to white and black, will be considered identical to its actual file, and will have the same hash.

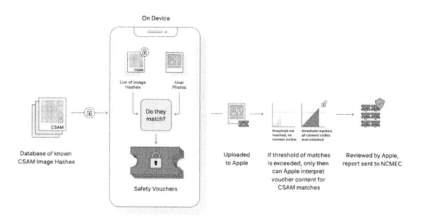

Apple said that, before a photo is saved into iCloud Photos, on-device harmonizing procedure is carried out for that picture against the encrypted array of recognized Child Sexual Abuse Material hashes. When there is a match, the device generates a cryptographic safety token. After a specific amount of tokens have been sent to the Photos and once a confidential threshold of matches is surpassed, Apple will then decode the tokens followed by a manual examination. When Child Sexual Abuse Material item is detected, the user account is deactivated and a signal is later dispatched to the National Center for Missing and Exploited Children. Apple will not share the entire threshold, however will make sure to an uttermost highest level of precision that accounts are not wrongly labeled. Some of usefulness of this Child Sexual Abuse Material detecting system includes; the system is very meticulous, with minute error rate of less than one in one trillion account per annual, being part of the procedure, and users also can't determine any bit about the array of well-defined Child Sexual Abuse Material photos that is utilized for harmonizing, so this secure the items of the database from mischievous use. In addition the framework is a fruitful means to recognize notable Child Sexual Abuse Material saved in iCloud Photos accounts while protecting user privacy, and much more.

Disabling Automatic Night Mode

Normally Night mode functionality is activated automatically when the camera sensor detects an outdoor or indoor surrounding with sufficient darkness that call for illumination, producing natural colors and noise reduction when using a camera of your device.
The Actual thing is that Night Mode wont at all times take the type of night mode photo that you desire.
 When you're desiring to shoot a natural evening background shot in which every origin of illumination stays inhibited. It is ideal to switch off Night Mode to avert extreme light exposure and overexposed shots. So if you don't want to utilize Night Mode while taking photos in low light, you can disable it by clicking the yellow Night Mode switch when it shows up at the upper-left of the viewfinder. However, the drawback is that when the camera app is launched

again and the sensor perceives low light, Night Mode will automatically be activated again. In the previous iOS updates you cannot turn off Night Mode as you can only manually disable Night Mode whenever the Camera app is open.

Thanks to Apple, now with iOS/iPad 15 you can deactivate Night Mode. And whether you reopen the camera app or not it will remain off until you switch it on.

To disable Night mode;

- Go to the **Settings** app on your device.
- Click Camera and select **Preserve Settings**.
- Turn on the Night Mode button. Green indicates it is on.
- You can still manually enable Night Mode in the Camera app regular methods.

Using Hide My Email

Another major privacy feature present in iCloud+ in iOS/iPad 15 is called Hide My Email. This feature allows you to generate special and random email addresses that automatically forward to an email account of your choice. What this does is that, you can send an email to someone through the random email address to prevent them from knowing your actual email address. With this feature you can hide your email address from some set people or a certain individual. This also helps to curb unsolicited mails from hitting your account.

Hide My Email is a handy measure to create a temporary email address that protects your original email address from spam messages. And it will also help you to recognize the organization that gives out your details when you begin to receive unwelcome emails from an email account you are sure is connected to one particular firm.

How to deactivate an email address for Hide My Email

Apart from creating a random email address, you can also deselect the created email address whenever you want. To do this;

- Open the Settings on your device.
- Select your Profile icon at the top of your display.
- Click iCloud and select Hide My Email.
- Select the email address you desire to deactivate.
- Click the option Deactivate email address.
- And hit the option Deactivate.

Henceforth, you won't get mails on the address again and to re-activate the address, go to the Inactive addresses section. Beneath the addresses list select the address you want to re-activate, and then click Reactivate Address.

Switching Your Forwarding Address for Hide My Email

You can swap your forwarding address, so that you can start forwarding all your random email addresses to a separate inbox. To do this;

- Open the Settings on your device.
- Select your **Profile icon** at the top of your display.
- Click iCloud and select **Hide My Email**.
- Toggle the option **Forward to**.
- Select the email you want to employ and click **Done** to confirm.

Personalizing Email Domain Name

You can create a custom email domain name in iOS/iPadOS 15, and then bring family members to utilize the same domain with their iCloud Mail accounts. So when your iCloud+ is activated, you can switch to a different iCloud Mail address without having to retain the iCloud domain reference. Also you can add family members to use the same domain name with their own iCloud Mail accounts. However it is unsure presently whether it will require the family members to join Family Sharing for this to fly.

Account Recovery Contacts for Apple ID

The Account Recovery Contacts for Apple ID is available in iPadOS and iOS 15, and enables you to select one or more reliable individuals to become an Account Recovery Contact. The implication is that the selected individuals can help you reset your password and recover your Apple ID account in case of a situation where you cannot access your account.

Digital Legacy Program for Apple ID

This is another Apple ID feature available to both iPadOS and iOS 15 powered devices. It is a kind of next of kin functionality for your Apple account. So with this Digital Legacy Program, you can appoint a family member as a Legacy Contact. Such that this person can access your account and personal details in case of your death.

Eye tracking support

With iOS/iPadOS 15 Apple continues to upgrade the accessibility features and one of the newly introduced accessibility functionality is called **Eye Tracking support**. Through this functionality the iPadOS devices uses a third-party MFi eye tracking hardware that monitors where you are looking on the display, making it possible for you to use your eyes to control your iPads. The cursor navigates to where you move your eyes to and can execute operations like selecting, clicking, swiping and scrolling.

How Enable Record App Activity

This feature helps to capture a 7-days overview of when apps access your location, microphone and detect if any third-party apps or websites have accessed your information. For this functionality to work you have to enable it first and turn on this feature;

- Open the Settings app on your device.
- Scroll down to Privacy.
- Tap on **Record App Activity.**
- Tap on the switch next **Record App Activity** to turn it on or off. Green indicates it is enabled.

At the bottom of the **Record App Activity** display under the Save App Activity you will see the summary there.

SECTION THREE

iPadOS 15 Multitasking

Apple made some upgrades to the iPad multitasking in these updates, you will still get the slide over and side by side apps functionality. There is a new-fashioned shelf functionality and multitasking menu at the top of any opened app interface to help you switch between side by side modes easily. On the top of each app you will see a new multitasking menu like three-dots and will give you three multitasking options when you click on it. Another thrilling additions to multitasking is called the Shelf and the multi-window shelf organizes all open windows of an app at the bottom of the screen for quick access. Just tap to exit or open switch a tab. The Split screen view is like separating your device's display into two halves and it can have the same or distinct apps interface together. For example you have opened Safari and you're reading some content on a website. To initiate side by side mode or split screen, drag up the app you want from the Dock to the left or right of the screen and then release it. This will give you the split screen view two apps. Slide Over is when an app is in full screen and you can launch another app or the same app to hover on the full display. Swipe on the bar at the bottom of the Slide over to switch between the stack of the slide over view. Tap and hold, then swipe up to close any of them. To go to the Multitasking Menu; launch an app and on the top center of the app's window click three dots. Then three or four options will pop up based on the app, which are full screen, split view, slide over and center window

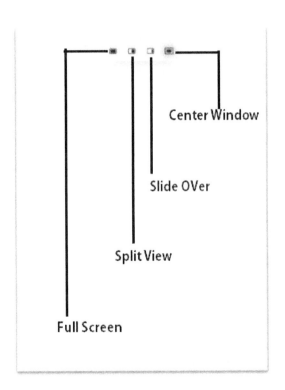

Center Window

Slide OVer

Split View

Full Screen

To split screen on your device, launch the first app you desire to work with in Split View, tap the three-dots on the top of the opened window and then click Split View icon on the Multitasking menu. The app interface will slide to one side of the display to give you space to launch another app from the Home screen or drag one from the Dock. And the two apps will open two halves on the display. There is a short vertical bar between the Split view, drag it the right or left to adjust the size of the view. Click on the three dots on the top of an app and then select the Full Screen icon to exit Split View. When in a Split View, the screen in which the three dots are highlighted is the active one. To access the Shelf, tap and hold on an app on the Dock and choose the option Show all Windows. Or click on the + icon to open another window of the app. Toggle on windows to navigate between them and swipe up on the display to exit it. Press the Globe key and Arrow down key to launch the Shelf from your Home screen. If you're in Split View or Slide Over and want to read a mail without having to close the view. Tap and hold on the email and pick the option Open in New Window from the small screen. And a centered window of the email will instantly pop up and user can access the mail and tap **Done** at the left corner of the window to quit. This center window functionality is only found in Messages, Mail and Notes.

What's New with CarPlay

Previously you could use your iOS device to unlock your car. But now in iOS 15 Carplay comes with an Ultra-wide band chip and it is designed to unlock and control your vehicle. Whether you have an iPhone or Apple Watch you can stop/start, unlock and control your car. Thanks to the U1 chip and its precision finding ability you won't be able to forget and lock your iPhone in your car. And the car will not start without your iPhone inside the car. However this will require some support from car manufacturers.

There is a new feature under Settings in the CarPlay called **Driving Focus**. Apple used to have Do Not Disturb in CarPlay, but it was replaced by Focus in iOS 13. And you have several pre-installed focus modes, which include Do Not Disturb, Driving, Sleep, Personal, and Work.

To enable CarPlay in your vehicle, go to the **Settings** app > **Driving Mode** and tap the switch next to **Activate With Carplay**. This feature will minimize the alerts you get while driving but some incoming calls and messages might still bypass this feature. Under the Appearance we have four new wallpapers and they both have a light and dark mode options.

What's New with Siri?

Siri is the personal virtual assistant that is designed by Apple and it was first introduced in iPhones 4S and later models. However, Siri is now seen in most Apple devices and it helps to automate work and give information. In iOS/iPadOS 15, Siri saw several significant upgrades and Apple brought in features that will offer an excellent user's experience. Apple devices running A12 chip or later, we have Siri that can feature on-device operations with offline requests support. With these updates, speech processing and personalization are achieved on-device, making Siri more secure and quicker at processing demands. The Apple Neural Engine helps to power the On-device speech processing and personalization. It is present on devices running with the A12 Bionic chip or later. On-device processing is available in English (UK, Australia, India, U.S, Canada.), Spanish (Mexico, U.S. Spain.), Japanese (Japan), Cantonese (Hong Kong), Mandarin Chinese (China mainland), German (Germany), and French (France). Majority of the audio requests made of are totally stored on the device and are no longer transferred to Apple's servers for processing. Siri's speech recognition and understanding of commands get better as a device is utilized, with Siri learning your choice topics, recently typed words, and most interacted contacts, with all of this data stored on device You can now share content such as web page, images, podcast or Apple Music song, podcast with someone. Simply ask to send it to that person and will be sent. Based on the presence of on-device processing, we have several requests that can be carried out offline. Sirican open apps, launch Settings options, set up and enable/disable alarms and timers, and control audio playback. According to Apple, also processes Phone requests, sharing, and Messages. Siri can now sustain context between voice requests. You can use Sirito control a HomeKit device at a particular time of the day. For instance if you want to turn on the sitting room lights by 8.p.m, you can say "Hey , turn on the sitting room lights at 8.p.m." This command also works for geolocation, so you can say "Hey , turn on the security lights when I am not at home. When is askedto control a HomeKitdevice this way, it sets up an automation in the Home app via the option Automation. To delete the Automation designed by Sirigo to the Home app. HomeKitdevelopers can also add support to their gadgets in /iPadOS 15. To use Siricommands with third-party devices you need a HomePod to route the requests through. With integration, you can use your device for commands such as, controlling devices, broadcasting messages, setting reminders, and many more. Before now, when using AirPods or Beats headphones Siri can announce incoming messages and calls but in iOS/iPadOS 15 Siri can now herald all notifications. Furthermore, can also announce reminders that show up when using or Beats headphones. Siri also will now announce incoming messages when is connected to a CarPlay setup. You have the option to deactivate/activate announcements when a message is read. And will not forget your choice and you can enable/disable this functionality in the Settings app. In iPadOS/iOS 15, the neural text-to-speech voices have expanded to Danish, Finnish, Norwegian and Swedish. It also came with Indic language and mixed English support. Can now execute requests in a native language and mix of Indian English, with, Telugu, Hindi, Marathi, Punjabi, Tamil, Gujarati, Bengali, Malayalam, and Kannada compatibility. You now have the option to bring in a section for **Suggestions** for websites you may desire to go to or content you might want to accessed via Start Page customization Beginning with iPadOS/ iOS 15, you will not be able to access in third-party apps to perform tasks such as booking taxis like Uber, make a payment, or generate fresh task lists on to-do apps. Most of these features may be switched by Shortcut options that can be triggered by voice commands. The Child Sexual Abuse Material guidance will be brought into Siri delivering more tools to assist parents, guidance and children get safe online experience. And further receive assistance with dangerous or harmful conditions. For instance, anyone who requests from Siri how to report CSAM or child exploitation will be directed to resources for where and how to log a report.

What's New with Notifications

The Notifications aspect of iOS/iPadOS 15 receives some major improvements in these latest updates as it now has a new look, flexible control and much more.

In iOS/iPadOS 15, app notifications now show huge app icons, making it more visible. Text messages from apps such as the Messages app will have the contact picture of the sender, thereby making it easy for you to recognize the sender. Because of the notification API this feature is compatible with built-in Apple apps and third-party apps. Therefore third-party apps like Facebook Messenger, WeChat, WhatsApp, and the rest can also display contact images in notifications.

In these latest updates, we have a feature called **Notification Summary** that helps you to control notifications from unimportant apps. And to know more about the feature, go to **Settings > Notifications >Notification Summary.**

Under **Notification Summary,** you can choose the definite times for alerts to be delivered. And also set which app notifications will be demoted to the Notification Summary. You can choose apps one by one, but your device might recommend the apps that send the most notifications. The Notification Summary can be designed to be delivered multiple times per day. The summaries will be set at 8:00 a.m. and 6:00 p.m. by default, however the times can be adjusted and more delivery times can be introduced. The Notification Summary also pops up on the iPhone in the allocated area before those times. Making sure you never miss any of your notifications both the irrelevant ones.

With these updates you can now momentarily muted apps sending notifications or Messages threads either for the next hour or for the whole day. When you're a member of a messaging thread that is very active but you aren't participating, your device will prompt you, recommending that you mute it to prevent disturbances.

Notifications that are time sensitive will still go through even with Notification Summary.

Apple designed a Time Sensitive notification API for developers for applications that require to deliver notifications promptly.

What's New with Photos App

This app was not left in the dark in iOS/iPadOS 15, as Apple introduces several much anticipated features. There is a customizable and accessible metadata info, new Memories functionality that allows users to fantasize their special moments, and Visual Lookup feature that enables you to recognize landmarks, pets, plants, books, etc. In addition there is a Live Text functionality through which text from pictures can be copied and pasted on your device. Apple restructured the Memories wing of the app in /iPadOS 15, making it easier than ever for your choice memories to reemerge. It comes with a new-fashioned look that comprises active cards with flexible titles, latest animation and transition styles, and numerous photo collages for additional cinematic user experience. You will see an entirely brand new memory view that is brought into Memories depending on each video and image, with Apple utilizing machine learning to administer the appropriate contrast and color customization for a steady view. The app features an interactive display that gives you the channel to click to pause, replay the last photo, skip to the next picture, with the music configured to continue playing while playing a Memory. Switching a song, adding or removing images, or tweaking a Memory look is carried out instantaneously without having to edit.

The new-fashioned Browse view makes Memories customization easy, to access every video and photos in a bird's eye view. The Photos app has consistently been integrated with Apple Music while generating and playing Memories, but with iOS/iPadOS 15 Apple Music people can bring in any song to a Memory. Customized song suggestions are generated by the Memories functionality depending on the individual songs impression and the content of his/her pictures and videos. The Memories functionality generates customized Music recommendations depending on an individual music preference and the content of your videos and images. Song recommendations might comprise songs that you played while you were traveling, songs that are trending at that time and site of the memory or a song from an artist when the memory contains a concert. To personalize a Memory swipe across several Memory mixes, which allows you to try out with individual songs, pacing, and Memory views. We have new types of Memories that are fashioned from your Photos library, with Apple bringing in new child-focused memories, international holidays and trends over time. The Pet Memories have been enhanced and iOS devices can identify cats and dogs. Via the Watch Next, when you're done with a Memory, Memories will recommend related content so that you can keep exploring your image content. The Photos app has the Suggest Less Often feature and what it does is that, when a person shows up continuously in your Featured Photos or Memories we can utilize the new-fashioned quick access functionality called Feature Less toggle to view less of that individual. And it also functions for places, dates and holidays. In iOS/iPadOS 15, text in photos on your device is now detected by the device via the all new Live Text menu. You can now highlight, copy and paste text in an image. And the text is also integrated with the Lookup feature, and it can also be translated. For instance, if the text in a photograph on your device is written in French, you can translate it to English or another language directly from the picture on your device. The Live Text feature is compatible with all the pictures in the app, live camera viewer, Safari and screenshots. Your device can identify different entities, artistic production, animals, books, plants, landmarks, etc. in specific images. For instance, when there is a picture of a flower on your device, via the Visual Lookup feature you can ascertain what it might be depending on a web search of photos that Apple automatically controls. When you see the **info icon** ⓘ with a small star over it while viewing a photo, it indicates that you have a Visual Lookup enabled. Click the **Info icon** ⓘ and then press on the little Heart icon to initiate a search. With iOS/iPadOS 15 running in your device, if tap the **info icon** ⓘ at the lower section of an image you will see the name of camera used in taking the shot, megapixels, shutter speed, aperture, lens type, size, etc. it encompasses every EXIF details you can access from the desktop app. Via the link Adjust you can bring in captions and set the time and date and all these are new to the iOS/iPadOS devices. According to Apple, with iOS/iPadOS 15 running in your device iCloud Photos syncs speedily when compared to previous iOS/iPadOS. Thereby allowing you to access your photo library swiftly. The people identification and naming was improved in iOS/iPadOS 15 and the

People album has boot recognition for the different individuals that are in your images. And the Photos app can identify someone in a great posture, people wearing gadgets, and people with closed faces. You also have the capacity to correct names mistakenly spelt Apple also introduced Photos image picker in these updates and it enables you to choose an image in a certain pattern for the aim of sharing. When you have an image chain, you can be confident that they're going to be shared in the flow that you desire. Through the **Shared With** You functionality in the Photos app you can access videos or images sent to you via the Messages app.   Images taken while you were around are displayed in the All Photos view and in the Days, Months, and Years view, as well as in Memories and Featured Photos. These pictures from the **Shared With You** menu can easily be stored in the Library. With /iPadOS 15in your device, you can search the whole Photos library via Spotlight. You can search by region/location, environment or scenes, people or content in the photos such as animals or pets via the Visual Lookup feature.

What's New with Messages App

To improve Messages user experience, Apple modernized numerous window elements and designed a brand new **Shared with You** functionality. The most significant development in iPadOS/iOS 15 Messages app is **Shared with You,** and it assembles things shared with users in Messages into the conventional apps. For instance you receive an image from an individual via the Messages app, it pops up in a **Shared with You** segment of the Photos app. Other apps harmonized with **Shared with You** includes Safari, Apple Music, Apple Podcasts, Apple TV app and Apple News. Every app that is compatible with **Shared with You** segments, items in these areas clarify the sender. You will get a name tag which you can click to lead straightaway to the Messages app to get a conversation of the thing you received. Clicking the name tag on the content you received also leads you directly to the related message in order for you to see the initial dialogue. When a very vital content like a link or an image is sent to you and you choose not to lose sight of it, you can press and hold on the content and then select the option Pin. And pinned content is displayed first in **Shared with You**, in Messages search, and the Details view of a Messages discussion. Numerous photos shared in the Messages app now pop up as a mini photo collage of images piled up on one another. Toggle the collage and swipe across it to view the collection. Tap on top left of the main window to view all the pictures in a grid view. And you will also see quick access tools to reply, download photos, add a tap back response, or share icons. Every picture you share via the Messages app has a small download icon next to them, which is used to save them. Thereby making it simply to save an image that you have forwarded to the photo library on your gadget. The Photos image picker is now supported in the Messages app, and it lets you choose pictures in a certain pattern for sharing. With iOS/iPadOS 15 the Messages app has nine new Memoji stickers such as shaka, lightbulb moment, heart hands, hand wave, etc. Apple added new Memoji, to ones used in the and Messages apps in /iPadOS 15 and they include 40 new outfit choices, new glasses options, new headwear options, a choice to select two separate eye colors and new accessibility options such as soft helmet, oxygen tubes and cochlear implants. Siri can now declare incoming messages in CarPlay. And when you are in Focus mode and an individual decides to mail you an iMessage, they'll get a status update prompting them that you've activated Focus mode. However, family and friends can bypass the Focus mode message in an event of emergency

What's New with FaceTime App

The FaceTime app in iOS/iPadOS 15 got several improvements, the FaceTime app comes with brand new features that swivel FaceTime into the apex platform for communicating with family, coworkers, staff, friends, loved, etc.

One of the major designs in FaceTime is called **SharePlay** and it is primarily a means to stay on FaceTime calls with your loved ones and friends.

From here you can listen to music, watch TV shows or movies together and share your screen.

During FaceTime call, you can decide to stream TV shows or movies and every person on the call will access the same synced playback and controls. Furthermore, you can view via iPad or iPhone, or switch over to the Apple TV to get a bigger display while still on call. As you discuss and watch, the volume will automatically adjust so that you can hear everyone talking without missing any moment.

To explore the multitasking experience of your device you can watch movies or TV shows, FaceTime and carry out other tasks.

While on FaceTime call with a group of people, if you play an Apple Music song, all members of the group will hear the song and can as well make an impact on the shared music queue. The SharePlay music screen shows synced playback controls, and the person in the group will see the next content. You can listen to music over your device , or send it over to the HomePod during FaceTime call.

In iOS/iPadOS 15, you can share your device's screen with a person on the FaceTime call, which is handy when you're choosing a movie, browsing through a photo album, planning a vacation, etc.

During FaceTime you can share content with the group via the group Messages chat.

Now in iOS/iPadOS 15 you have the ability to create a FaceTime link, through which other persons can join your FaceTime call. The links give you the avenue to schedule FaceTime calls ahead of time. And you can then send the links to other individuals so that they can join the meeting or conversation at the right time.

FaceTime fuse instantly with the Apple Calendar app. And for you to generate a link, launch the **FaceTime app** > **Create Link**. And the link can be shared via an email, text message, AirDrop, etc. And you can join the conversation by tapping the link.

As result of the link creation functionality in FaceTime , you can now join FaceTime calls on the web, indicating Android and PC owners can now take part in FaceTime calls.

However, the link can only be created via Mac, iPad or iPhone and once the link is created anyone can then join the meeting. Use Chrome or Edge browsers when joining from the web and FaceTime from the web works one-on-one or in Group FaceTime calls.

FaceTime now has Spatial Audio support in iOS/iPadOS 15, so if you're in a meeting with a group of individuals and their images are in a distinct position on your display. Their voices will sound like they're approaching from the appropriate point on your device. To explore Spatial Audio you will need an iPhone or iPad with the A12 Bionic or later.

The new Grid View functionality in FaceTime has placed the app on the same level with other video conferencing apps. You can arrange every member of the group into the same-size tiles, and the person talking is automatically selected.

Like in every other camera setting, Portrait Mode in FaceTime blurs out the background in your back and focuses on you. Also for this functionality to work you will have a new device with A12 Bionic chip or later.

When speaking during FaceTime call, if you're muted, your device will notify you that the mute toggle is enabled.

If you're making use of the rear-facing camera on a FaceTime call, you have an option to zoom in on your display.

We have two microphone modes in iOS/iPadOS 15. The first one is the **Wide Spectrum** mode designed to ensure the ambient noise is audible, which is perfect for group conversation. And the second one is the **Voice Isolation**, which is configured to reduce background noises to prioritize your voice.

What's New with Maps App

In iOS/iPadOS 15, the Maps app also saw numerous improvements and now has upgraded transit directions, additional deep AR-based walking directions, and superior driving directions. The Maps app now offers comprehensive road information when utilizing the app for driving directions. It will make lanes, crosswalks, bike lanes, and turns plain to you, so that you can be sure of your destination. When you come across a multiplex interchange that requires to be steered, Maps will reposition into a street-level viewpoint to reduce the uncertainty. You can see these in a well-designed driving map that also features the latest traffic happenings and circumstances such as shutting down of road. And you will be aware of what might occur on your path. You have Transit, Driving, Satellite, and Explore maps to select from. And we have a new window for selecting a map, with previews of the view of every map. With /iPadOS 15, Maps is comprehensive in every area, however Apple hand-picked some cities and stretched the data extra mile. Neighborhoods, roads, buildings and trees are displayed in depth detail, and there are 3D representations of landmarks. Examples of Cities with 3D landmarks and high levels of detail are London, New York, San Francisco, and Los Angeles. Apple has also displayed its Infinite Loop and Apple Park campuses in 3D, and this functionality will spread out to more cities in the future. Apple Maps now features dark/night mode with new colors and a modernized look, and commercial districts are selected. You now have walk directions in augmented reality, making it stress-free to reach your destination, most especially in conditions in which the directions are complex. The AR mode provides bit-by bit directions when walking, and you can switch the mode by scanning the structures around you when alerted after starting a route. To access AR walking directions, you need a device with an A12 chip or later. In these updates, if you keep zooming out until you can no longer zoom further in Maps, a globe view will come up that enables you revolve around the universe and drill into fresh locations. In order to boost this globe view, Apple introduced extra details for ocean, deserts, mountain ranges, and forests. iPadOS/iOS now has a functionality, which enables you to set a time to arrive or a time to leave when receiving directions in the Maps app, thereby letting you to obtain probable commuting lengths before time. In metropolises with improved maps, the transit map has been boosted with extra detailed view that displays popular bus routes. Regular transit riders can view the whole of the departures that are filed close to them and pin their preferred lines. So that the lines can always pop up at the top of the Maps interface. In public transit, you have an enhanced user window that simplifies your view and how you communicate with the routes in a one-handed functional system, and it's handy when there is no vacant seat at hand. As the stop comes closer on public transit, Maps will notify you that it's almost time to get off. You have a dedicated Guides Home that contains editorially curated guides with points on how to go about the city where you stay or the one you're visiting. The Search functionality in iOS/iPadOS 15 Maps has been upgraded and search results can now be sieved by options like specific cuisine, whether a restaurant is open, whether a restaurant offers takeout, etc. when you hover on the map after performing a search, the Maps app will automatically update the search results to the current region. With iOS/iPadOS 15 Maps has a user account which accommodates Maps preferences such as direction mode, toll, etc. It also builds up Photos, Guides, Favorites, and Ratings. Maps will now prompt you when there are weather set up that will serve as a barrier on your pathway. For instance Maps will reroute you within the flooding when there is one on your path. Alternatively, the app can alert you as regards the barrier, so that you can circumvent it. And place cards for restaurants, businesses, landmarks, etc. got a major boost with a new structure. Physical features such as mountain ranges have its dedicated place cards alongside data such as elevation.

What's New with Find My

The Find My app has continued to enjoy Apple attention ever since it was introduced into the iOS/iPadOS devices. This time in iOS/iPadOS 15, **Find My** also got numerous upgrades and some new additions to the app includes live locations and the capacity to keep tabs on your device even when it is shut down or after it has been erased.

Via the Live Locations feature, when you monitor spouse, offspring or friends with the **Find My** . It will now display continuous streaming updates of their new position rather than notifying with a new location every few minutes. This Live Location functionality was configured to deliver a prompt impression of direction, speed and advancement when you're viewing an individual's location. You can still track devices that are off via the Find My network in iOS/iPadOS 15. Regardless of the reason for turning off the device it can still be located when it is near to another Apple device. The Find My network helps to detect Apple devices without a Wi-Fi or cellular network connection by depending on other nearby Macs, iPads and iPhones. The U1 chip, NFC, or Bluetooth continues to be running in the background even when your device is shut down or when the battery is low. However, in the case of low battery, tracking may last a few hours. You have to turn on the Find My network feature for it to work. Though the feature is enabled by default, you can check to be sure via the profile page by clicking on Find My , selecting Find My iPhone, iPad, or Mac, and then ensure **Find My network** is turned on.

When your device is stolen and wiped out, with iOS/iPad 15, it can still be traceable and will pop up in the Find My app. This feature is secure by Activation Lock that prevents another from using your device without producing Apple ID and password.

In the previous iOS/iPadOS updates, wiping a device would still make Activation Lock active, so no person can access your device without your password. However, wiping a device will stop the Find My from working. That is, the erased device won't pop up in the Find My app and that is not how it works now in iOS/iPadOS 15. If Activation Lock is active and the device was not erased using your username and password to deactivate Activation Lock. Then the Lock is attached to your account and traceable via Find My . The Hello screen will indicate that the device is locked, traceable using Find My and it belongs to another person. This will prevent the theft from selling off the device.

With Separation Alerts in iOS/iPadOS 15, Find My will notify one of the other devices with you when an iPad or iPhone is left behind. This feature works with AirTags and third-party gadgets that are incorporated with the Find My network. For instance, if you leave a MacBook or wallet behind at your office after you left with your iPhone, your phone will prompt you.

The feature Separation Alerts can be enabled in the Find My app and you have to do it for every individual item that you possess. And you can also configure exceptions, so when you forget something at home on purpose, you won't get any notification.

The Find My network can track AirPods Pro or AirPods Max location if misplaced.

With the Find My network functionality in iOS/iPadOS 15, AirPods Max or AirPods Pro could interact with other people's Apple devices, enabling you to track them down when they're out of range of your devices.

You now have a Find My widget in iOS/iPadOS 15, that can be introduced into the Home screen or the Today View to monitor items quickly without having to go to the Find My app.

The Widgets are available in two sizes medium and small, and can be used to keep tabs on items or people. The People feature focuses on your friends and family members who have shared their location with you. And the items widget tracks are AirTag items and third-party gadgets that use the Find My network. Clicking any of the Find My widget launches the whole Find My app.

Majority of these new features in Find My app are restricted to the iPhone, however, iPad compatible with Separation Alerts, the Find My Widget, and live locations for family and friends.

Macs with the latest macOS Monterey support the Find My widget and for live locations for family and friends.

Nevertheless, Offline tracking and tracking for wiped out devices are features exclusively for iPhone.

What's New with Wallet App

One of the significant features to the Wallet app in iOS/iPadOS 15 is **Digital IDs.** Here in the United States, Apple is liaising with authorities to let iPad or iPhone users link their state ID card or driver's license to the Wallet app. there by making digital IDs to be utilized in place of a physical ID card. Also, Apple is collaborating with the Transportation Security Administration (TSA) to let their checkpoints support digital ID cards. So with digital ID cards you can board a plane.

And with the latest iOS/iPadOS updates ID cards saved on your device can also move to Apple Watch.

According to Apple, digital ID cards will first be available in some states here in the United States and let's keep our fingers crossed as we wait for Apple to unveil the states.

However, each state and Apple have to come to terms before the Wallet app will get the green light to serve as state ID card or driver's license storage.

Also, Apple is planning to switch the conventional physical keys with digital forms, which you can access directly from the Wallet app on your device. However, this key features will need Apple and the other relevant firm to be on the same page before it will work.

Now in iOS/iPadOS 15, corporations that design HomeKit-supported locks can enable Wallet-based home keys, which can be employed to click to open HomeKit door locks. The Wallet app will house the Home keys, which are accessible via your device and Apple Watch.

In addition, Organizations can enact digital keys for the Wallet app, making its staff open office doors just by clicking on their device in place of using the firm's access card. Also hotels that have support for digital keys can also give access to their client to add those keys to their Wallet app after securing a reservation, thereby freeing them from the Receptionist documentation. You can use the digital hotel key stored in the Wallet app on your device to open the hotel room. But when you sign out of the hostel, the room key is cataloged automatically.

Now with iOS/iPadOS 15, Safari allows users to download multiple passes to the Wallet app with a click. So when you purchase a TV show, movie or zoo passes for family members or friends, you can download them to your Wallet app by a click instead of downloading each one at a time.

The **Expired** tab of the Wallet app accommodates every obsolete event ticket and boarding pass.

The **Wallet & Apple Pay** menu in the Settings app has a tab called **Hide Expired Passes** is an option in the Wallet & Apple Pay section of the Settings app. And this can be disabled if you desire to hold on to your passes, however the feature is enabled by default.

In iOS/iPadOS 15, we have Ultra Wideband support for accurate spatial awareness. This enhanced spatial awareness enables you not to forget your device in a car and it also prevents firing of the ignition of a car without having device in the car. The updates also come with support for preheating the car, locking or unlocking the car, honking the horn or opening the trunk with control alternatives that are situated in the Wallet app when you're close to your car.

Currently, the Bayerische Motoren Werke GmbH (BMW) is the sole vehicle producer that has enacted Car Keys support, so your car maker have to support this feature for it to run in your vehicle.

What's New with Weather App?

The Weather app received significant upgrades in iOS/iPadOS 15 and the much anticipated Dark Sky functionality has been integrated into the app. Now you can easily comprehend information from the app at a glance. The app has taken on the card label window the same as that of the Settings app, splitting various data into categories.

Apple kept the main view to house weather conditions on an hourly basis, which you can swipe across. Also we have a new 10-day forecast interface that allows you to get what weather you can bank on in the future. With the improved 10-day forecast you will see anticipated weather conditions with a bar that displays the temperature against time at a glance.

At the lower section of the app we have a new weather module, through which you can add new graphical weather data. We have modules for humidity, pressure, temperature, sunset and sunrise, UV index, precipitation, wind, visibility, and air quality. And every one of these modules comes with context, graphics and additional information.

The app has a full-display weather map that features the summary of anticipated temperature, air quality and precipitation. To access maps, toggle on the default temperature map and click on the stack to switch the display to air quality or precipitation. Also you can access the weather maps from any interface via a little folded map icon at the bottom left corner of the app. The temperature and air quality maps allow users to access the weather conditions of their location and neighboring area. And you can obtain air quality data in the United States, Mexico, Canada, South Korea, China-mainland, India, Italy, Spain, Netherlands, UK, France and Germany.

However, the precipitation maps are animated and they reveal the trail of incoming storms and the intensity of rain and snow. To have some lead when it is going to hail, rain or snow, register for next-hour precipitation alerts and this feature is only obtainable in the United States, Ireland and UK.

The Weather app in iOS/iPadOS 15 comes with tons of animated backgrounds that deliver additional details about rain, storms, sun position, clouds and much more. The animated backgrounds switch all over the day, in the night and based on weather influence.

What's New with Translate App

The Translation functionality of iPhones and iPads was not left behind in the latest updates. With iOS/iPadOS 15 the Translate app features Live Text translation, Language Selection Improvements, Auto Translate and much more.

Now because of the enhancement of language selection functionality, you can easily choose languages via drop-down menus.

The Translation functionality offers auto translate speech feature without having to press the microphone button while you're talking. The system will automatically know when conversation starts and ends, this will enable the other individual to give you feedback without having to tap his/her device.

Now in iOS/iPadOS 15 the conversation view in Translation functionality comes with a face to face feature. Such that every individual in the conservation will see their own side of the chat.

The Live Text functionality in the translate app enables a device to pick up text in any picture or image on the device. You also have the ability to copy and paste text, and also utilize the built-in system-wide translation functionality to translate text. For instance you travelled out of your country and you don't understand the language of the region where you're. You can simply take a shot of the platform where the text is written. Then select the text, and tap the translate option to get the result.

Live Text functionality is supported in screenshots, live previews with the Camera app, Safari, Quick Look, etc.

It is now easy to switch to the conversation mode. Simply click the Conversation tab in landscape or portrait view at the bottom of the Translate app.

What's New with Files app

The Files app in iOS/iPadOS 15 got some major innovations and of them is the support for NTFS drive, but it is read only and limited to the iPadOS.

So with iPadOS 15 running in your iPad you can access any NTFS files but cannot write to NTFS volumes. Insert your hard drive, SSD, or flash drive and import any files you want.

Also the Files app comes with drag and drop support, which is handy when working in a split view. So you can now drag and drop photos, documents and files from one app to another.

Before now when you're transferring a file through the Files app, you won't know how far the operation is going. But with iOS/iPadOS 15 when transferring a file via the Files app you will get a mini circle bar at the top-right corner of the screen. And if you click on the mini circle bar it will pop up a progress bar. The progress bar has an **Edit** button to cancel many transfers by a tap, you also see the amount of time to complete the operation and size of the file.

You can swipe to the left on the progress bar to access a cancel button. With this new feature you can cancel transfers you mistakenly initiated.

With a Magic Keyboard, mouse or trackpad connected to your iPad you can click and then drag to form a marque to select multiple folders or files based on your preference. And this functionality comes in handy when dealing with files or folders.

Another addition to the Files app is the Groups functionality and it organizes your files based on certain properties. And you can group your files according to their kind, date and size.

Meet Siri

Just like we have Cortana from Microsoft, Alexa from Amazon, Bixby from Samsung, and Google Assistant from Google, Siri is also a virtual assistant designed by Apple for her devices. And you will basically get Siri if you have any of Apple devices like Mac, iPad, iPhone, HomePod, Apple TV and Apple Watch.

Siri can execute several tasks for you, ranging from inquiry about time, weather condition, initiate a call to News updates, etc. Also the assistant can turn off or on settings, search for items, create reminders and alarms, send a message and many more.

To trigger Siri on your iPad or iPhone, tap and hold the Side button or Home button depending on the button on your device. With the recent iOS/iPadOS updates Siri will show up as a little icon at the bottom of the device's screen.

Majority of the Apple devices come with the Hey Siri activation command support, however most recent Macs, Apple Watches, iPads and iPhones feature the hands-free Hey Siri support when linked to power or not. This indicates that the Hey Siri activation command can be employed to trigger Siri whenever you want.

In a scenario where several devices that would react to Hey Siri commands are present, Bluetooth functionality will be used to ascertain which of the devices that will respond to the command to prevent all of them from responding at the same time.

Also you can interact with Siri and because the virtual assistant is incorporated into other facets of iOS and watchOS.And it can perform dynamic recommendations, which you can respond to. In WatchOS, iOS and iPadOS devices Siri can execute several form suggestions. For instance, if you plan to attend a meeting with business, Siri could suggest that you remind the person via call on the Home screen or when you navigate to find and see the Siri recommendations options.

Siri can pop up search suggestions in Safari, and in the Mail and Messages app it can recommend email addresses or telephone contacts depending on your entry.

Other suggestions from Siri are; time to leave if you set up a schedule-, upcoming programs on your calendar, etc. Note this recommendation from Siri is not determined by the system rather it is based on how you use your device.

We also have Siri Shortcuts functionality, and it is a shortcuts and automation feature that enables you to carry out tons of operation of your device.

Siri is available in most countries from Asia down to Africa and they includes, New Zealand, Australia, Mexico, Austria, Norway, Belgium, United States, Brazil, Canada, Chile, China, Finland, South Korea, France, Germany, Switzerland, Hong Kong, India, Saudi Arabia, Ireland, Israel, Sweden, Italy, Japan, South Africa, Malaysia, United Kingdom, Russia, Singapore, United Arab Emirates, Spain and much more

And some features like sports info, translations, conversions, calculations, restaurant information, showtimes, reservations, dictionary, and movie information are only available in countries.

Some of the requests and questions Siri can execute are; Sending/reading texts, Set reminders, Control HomeKit products, Triggering calls/ FaceTime, Set alarms/timers, checking calendar, updating sports scores, Search and create Notes, Making reservations, launching and interact with apps, Solving mathematics equations, translations, conversions, Activate Siri Shortcuts, Checking the weather and many more

SECTION FOUR

How to Blur Background during FaceTime Call

The iPadOS/iOS 15 has numerous FaceTime improvements like new audio and visual effects to boost the video call experience. And one of the new visual features is Portrait orientation while making calls. When this Portrait orientation is activated, the caller's background can be blurred so that he/she will not be concerned about what is behind them. Video conferencing software like Zoom, Microsoft Teams has this ability to show a different background, thereby saving you from distractions or embarrassment.

Now to blur your background during FaceTime calls;

- Open the **FaceTime app** and then start a video call.
- Go to the **Control Center** with a diagonal swipe down from the upper-right corner of your display.
- And toggle the **Video Effects** button.
- Then swipe up from the bottom of the screen to switch to the call screen.

There will not be much inconsistent in your looks when the Portrait orientation is active except you open up the view of yourself. However the individual on the other side of the call will see your domain vaguely blurred, like it used to appear in Portrait images.

Hindering Emails from Tracking Users with Mail Privacy Protection

The iPadOS/iOS 15 offers users a new email privacy feature known as **Mail Privacy Protection**. This feature helps to block advertisers and big organizations from tracing the way you communicate with emails from them.

With App Tracking Transparency feature you can disengage from hidden tracking that third-party apps conventionally use to promote ads. Though your email inbox can still be under watch.

Often when you open unsolicited marketing emails from a firm they are aware of the time you open it. With tracking channels used by the firm they can as well know your location at the time you opened the email. Most of this stalking is accelerated by remote images that pops up when opening an email, and at most are crafty with advertisers using unseen tracking pixels. Once the mail is unveiled in your email client, the code inside the pixel quietly conveys link details like your IP address back to the advertising firm.

So what the **Mail Privacy Protection** features does is to mask your IP address and load all remote data confidentiality in the background, dispatching it via several proxy platforms and aimlessly allocating an IP address. Rather than getting your IP address they will see the one that correlates with your location. So this will deliver indefinite details of you that are not exact and cannot be used to get your precise location and data.

And when you launch the Mail for the first time on your device running iPadOS / iOS 15 you'll be prompted to turn on privacy protection and note that the feature is not activated by default.

However you can disable or enable Mail Privacy Protection manually and to do this;

- Go to the **Settings app** on your device.
- From the **Messages** option click **Privacy Protection**.
- Click the switch next to **Protect Mail Activity** to turn it **Off** or **On**.

Despite turning on this feature, advertising companies can still keep tabs on your profile with tracked connections when you finally open their emails. So enabling **Privacy Protection** is not enough, don't even click the emails. However, unexpected behind the scenes tracking is now a thing of the past.

Hiding App Notification Badges on the Home Screen

You can now conceal or filter incoming notifications, consisting of notification badges for every app on an iPad or iPhone Home Screen. And to hide the notification badges on device;

- Pull down the **Control Center** by swiping down diagonally from the upper right corner of the display.
- Click the **Focus** button.
- Toggle the ellipsis ⬤ button on the top right corner of the small screen.
- And click select **Settings** the dropdown.
- Under the heading **Customization** click **Options**.
- Hit the switch next to **Hide Notification Badges** to activate or disable the feature.
- Go ahead to the **Control Center** once more, click **Focus**, and then activate the Focus mode that you just selected to hide notification badges.

When Focus mode is activated, it syncs across Apple devices, such as iPhone, Mac, and iPad.

Correcting Errors in the Photos People Album

The People album in the Photos app helps to detect faces in your photos in order to tag people's individuals with their real names and use that as a tool to categorize in your library tab.

For the regular Apple device owner, you will notice that the built-in face recognition in Apple devices is not void of errors. As in often times there will be some person's face that will be labeled wrongly.

So with iOS/iPadOS 15 running in your device, the People album has the much enhanced ability to detect numerous faces of persons in your pictures.

Also, with iOS/iPadOS 15 you can make corrections to wrongly labeled photos and to do this;

- Go to the **Photos** app on your device.
- And then select the **Albums** tab.
- Under **People & Places** click the option **People**.
- And choose a subject or an individual.
- Click the **Ellipsis** icon at the upper-right corner of the display.
- Highlight the option **Manage Tagged Photos**.
- Click to deselect images that have been wrongly labeled.
- Toggle option Tag More Photos at the bottom of the screen brings untagged photos of the individual.
- Click **Done** at the upper- right to confirm.

Viewing EXIF Metadata in the Photos App

With iOS/iPadOS 15, the Photos app was upgraded with an expanded Info pane where you can view details of an image in your library. The details of the photo you will get including EXIF metadata like name of the camera, the shutter speed and lens type. Also via Info pane you will see the file size of the image, and the file's location if it was saved from within another app.

To access new information pane in the Photos app a device powered by iOS 15/iPadOS 15;

- Launch the **Photos app** and toggle a photo in the Library tab.
- Click the **info button** ⓘ at the bottom of the screen just beneath the photo display.
- And search for the EXIF data in the box beneath the date and time.

Via the link **Adjust** you can edit the images time of shooting and the rest.

Choosing Images That Shows Up in a Photos Memory

With iOS/iPadOS 15, users can adjust different memories to deliver excellent experience and interaction. One of the significant changes in the

Photos app is the refined Memories. And it happens to be the means to effortlessly organize the images that should show up in a memory. With the new interface, you can quickly select which pictures you want in the memory and which ones you want to leave out, and your changes will be remembered next time.

Adding Apple Music Songs to Your Memories in Photos

The iOS/iPadOS 15 Photos app comes with some noteworthy updates to the Memories tab and they include a new-fashioned design, better collaborative window, present-day visual effects, and loaded integration with Apple Music. When framing and watching Memories, the Photos app usually supports basic music integration all the time. However with the latest updates users can decide to bring in any song to a memory. When a memory is picked in the app, new Memory Mixes will put forward songs based on your Music listening history which may progress with the videos and pictures. Alternatively, you can decide to hand pick a song for a better individualized experience. The Memory mixes feature can also recommend Memory Looks which are basically color filters. And these recommendations might flow with selected memory and partnering music so as to induce a specific ambience. To add songs from Music your Memories in /iOS 15Photos app;

- Open the Photos app on your device, and go to the For You tab at the bottom of the display.
- Under the Memories option, choose the memory like to customize.
- Click the playing memory to display the controls overlay.
- Toggle the Memory Mixes icon at the bottom left corner.
- Swipe right or left, and via machine learning the Photos app will unite recommended music tracks from Apple Music with distinct Memory Looks that might connect.
- Tap the **Restart** icon at the top left corner to rewind the memory. When you see a song you love, just hit the display, then the song and look will sync to your memory.
- Or toggle the Music icon with the + **sign** to pick another track.
- Go to the Search icon at the top right to get a particular song in your music library or browse Music's Top Suggested songs and other categories.
- Click **Done** after selecting a song to preview it in your memory.
- And tap the display and the music will be sync to your memory.

Apart from integration of Apple Music into the Photos app, the app also has new Memory types and they are new international holidays, child-focused memories, and trends over time, and upgraded pet memories with the potential to discern separate cats and dogs

Improving Photos App Memories with 'Memory Looks

With iOS/iPadOS 15 you can personalize the appearance of videos and images that are present in the app's auto-generated Memories. **Memory Looks** are usually video/image filters and can bring an outstanding humor to the items of Memories.

We have 12 Memory Looks that you can explore and every look operates by analyzing every video and image. By using the right proportion of contrast and color adjustment you will always get a great look.

And to enhance looks in your device's Photos app;

- Open the Photos app on your device.
- Go to the **For You** tab at the bottom of the display.
- Under the **Memories** option, choose the memory you want to edit.
- Click the playing Memory and pause it.
- And select the **Memory Mixes** icon with a star at the bottom left corner.
- And then toggle the **Memory Looks** icon at the bottom right corner.

- Hit a Memory Look thumbnail preview to choose it.
- And then toggle **Done** at the top right corner to confirm.

How to Copy and Paste Text from Photos

With iPadOS / iOS 15 Live Text feature you can identify text in an image you captured or the on your device viewfinder as well as make some adjustment on the image. For instance, you could snap a billboard showing the contacts such as email and telephone number of a company. So you can extract the number to call or save. Also, you can copy Live Text from images.

To do this;
- Open the **Photos** app on your device.
- Search for a **photo** with some text, like online market interface or an office address. And **zoom out** by **Pinching** on the picture to make the words more visible if they were too small for you to see.
- **Long – press** on the words, and then drag the ends of the selection tool to highlight all the text that you desire to copy.
- Then remove your finger from the display and go for **Copy** from the small menu that will show up at the top of the display.

- Navigate to an app which supports text input, long-press where the cursor is situated and click **Paste** from the displayed menu.

Translating Text in Photos

In iOS/iPadOS 15, the translation app took advantage of the new Live Text feature to translate other languages that show up in a picture.

And you can also translate text whenever it shows up in the viewfinders of Apple's stock Camera app.

- Go to the Photos app on your device.
- Select a photo saved in the library with words in a language that you want to translate.
- Click the **text capture icon** that appears in the lower right corner of the picture.
- Toggle the desired section of the captured text and drag the selection tool to highlight the text that you desire to translate.
- Click the **right arrow icon** in the small menu that is displayed at the top of the screen to access more options.
- And then click the **Translate** from the small menu that is displayed on the screen.

A panel will spring out from the bottom of the display displaying the translation beneath the words selected.

Also, you can decide to copy the translation, change the translation to another language, and hear the translation pronounced via the **Actions menu** that shows up when paste it in another interface.

Using Spotlight to Search for Pictures

With the iOS/iPadOS 15 the Spotlight Search can search for photos in your device. The Spotlight Search is now incorporated with the Photos app and other numerous ones.

To access this feature;

- Launch the Spotlight Search by swiping down on the **Home Screen** or **Lock Screen**.
- Enter "Photos," then start searching your images via individual locations such as people, scenes, or even things in your photos like plants or pets, thanks to Visual Lookup.

You can also see images in the search results as Suggestions. For instance if you enter dogs, you'll get your pictures showing up with results from the Files app, web, Siri, etc.

However, if you don't want your photos popping up in the Spotlight Search, go to the **Settings app** > **Siri & Search** > **Photos**. From the Photos window you can enable which apps have access to your photos.

Setting Arrival & Leaving Times for Driving Directions in Apple Maps

You will get a much improved Apple Maps in iOS/iPadOS 15 and among new features are, interactive globe, improved driving directions and new comprehensive info of cities. Apple also added the potential to set leaving and arrival times for driving directions to the Maps app.

To set up the leaving and arrival times for driving;

- Open the **Maps app** on your device.
- Go to the **Search** bar, type your destination and then hit **Search**.
- Click the **Driving directions** button normally in meters.
- Toggle **Leaving Now**.
- At the top of the screen via the **Leave** at and **Arrive** option, choose a time and date.
- Click **Done** at the top right of the confirm to confirm.

So, with these updates Apple Maps app can access many directions and time taken to reach your destination depending on speculated traffic for the set time and date.

How to Drag and Drop Screenshots

Cross-app drag and drop functionality has been present in iPadOS devices for some time now and with iOS 15, iPhone now supports drag and drop functionality across apps . Making files, photos and text movement from one interface to another easy.

Apart from apps you can aso have the multi-finger functionality support with screenshots. And to do this;

- Snap the screenshot by pressing the Side button and the Volume Up button together.
- Long-press on the screenshot thumbnail in the lower-left corner of the display, and stand by for a few seconds for the white frame around it to vanish from your sight.
- Click the app that you wish to use the screenshot in.
- Go to the location where you want to use the screenshot and choose the particular album you generated.
- Take the screenshot into the desired location and release your finger to drop it.

Turning On Wide Spectrum Audio on FaceTime Call

Apart from the new visual and audio effects improvements in FaceTime video and audio calling interface in iOS/iPadOS 15, you will also get a Wide Spectrum mode audio feature. The Wide Spectrum mode is a microphone mode that ushers all the available sound into a call, making it possible for the other individual to pick up your background sound.

And to activate Wide Spectrum mode;

- Open the FaceTime app on your device.
- Then start a video call. Swipe down from the top-right corner of your display to go to the Control Center.
- At top-right of the Control Center panel click on Mic Mode toggle.
- From the small panel that will show up select Wide Spectrum to turn it on.
- Close the Control Center and back to the call interface by swiping up from the bottom of the display.

If you don't want the other person to hear the sound of your background, you can turn off Wide Spectrum mode by selecting **Voice Isolation** mode or **Standard**.

Inviting Android Users to a FaceTime Call

With iOS/iPadOS 15, someone without an iPhone, iPad or Mac can join a FaceTime call via a link to the FaceTime conversation that you created.

The new FaceTime Links feature enables people who don't have an Apple account to sign into a FaceTime call through a web browser opened on any non-Apple device such an Android smartphone or Windows computer.

However, the person who will initiate the FaceTime call, create and share the FaceTime link must have an iOS, iPadOS, or MacOS device.
To do this;

- Open the **FaceTime** app on your device.
- Click on **Create Link**.
- Name your FaceTime Link via the link **Add Name** under FaceTime Link.
- Choose the channel of sharing your link from the Actions menu like Messages, Mail, Notes, AirDrop, etc.

When the other individual gets the link and clicks it, they'll be directed to a web page where they can type their name to join the conversation. When they have successfully joined the call, they'll have the regular FaceTime features turn off video, change the camera view, leave the call and mute/unmute the microphone.

Blocking Background Noise in FaceTime Calls

One of the new features of FaceTime in iOS/iPadOS 15 is **Voice Isolation**, which can also be handy with third-party apps such as Teams and WhatsApp to deliver a better video call experience.

Normally during FaceTime calls, the device's microphone catches a vast range of sounds in the surrounding area. However with the Voice Isolation feature, the device uses machine learning to pick out these sounds. And then hold off every ambient noise and enhance the caller's voice to deliver an excellent experience.

And to block background noise on your device during FaceTime calls;

- Open the **FaceTime** app on your device.
- Start a FaceTime video call or use a third-party app with a video calling feature.
- Swipe down from the top-right corner of your display to go to the **Control Center**.
- At top-right of the **Control Center** panel click on **Mic Mode** toggle.
- From the small panel that will pop up select **Voice Isolation** to turn it on.
- Close the Control Center and back to the call interface by swiping up from the bottom of the display.

Rearranging and Deleting Home Screen Pages

With iOS/iPadOS 14 you can disable individual Home Screen pages with the help of App Library. But not everyone likes the App Library, and it's not possible to rearrange the order of your Home Screen pages, or delete a page straightway.

But in iOS/iPadOS 15 you can delete and rearrange your home screen pages and to do this;

- Long –press on an empty space on the Home Screen to switch jiggling mode.
- Click the row of dots just above the Dock showing your Home Screen pages.
- From the Home Screen grid displayed on your screen, tap and drag a page to rearrange it relative to the rest of the pages. The rest pages will switch position to pave way in response to your dragging action.
- Click **Done** in the top-right corner of the screen when you are finished with the arrangement.
- And again click **Done** to quit jiggle mode.

On the other hand, to delete home screen pages;

- Long –press on an empty space on the Home Screen to switch jiggling mode.
- Click the row of dots just above the Dock showing your Home Screen pages.
- From the Home Screen grid displayed on your screen, click the checkmark under the page that you desire to delete.
- Toggle the minus (-) icon at the top-left corner of the page to delete it.
- Click **Done** at the top-right corner of the display after you have finished.
- And tap **Done** again to quit jiggle mode.

Note that, if you delete a Home Screen page, the apps that are on the page are not wiped out but are still available in the App Library. Drag the apps from the App Library when you want to bring them back to the Home Screen again.

Getting AR Walking Directions in Maps

In iOS/iPadOS 15, Apple Maps app saw huge improvements and new features, like the potential to utilize augmented reality walking directions in major metropolises. The latest AR mode can map walking directions upon the real world via your device's backend camera, thereby providing means to view your destination in built-up areas without having to check your device frequently as you progress.

First and foremost, start a walking route, then lift up your device and scan the structures in the neighborhood when notified. And a step-by-step guidance will automatically be displayed in the AR mode, thereby making it easy for you to access your destination.

To have access to this feature your device should have the A12 chip or later and iPhone or iPad released from 2018 to this present day are compatible.

How to Share Display Using Siri

With iOS/iPadOS 15 running in your device you have access to improved on device processing and offline requests features in Siri. The voice assistant appears more context-aware – which allows you to share any content such as images, songs in Apple Music, website, etc. displaying on your device screen with another person through the Messaging app. Simply say **Hey Siri,** and then say **Share this with (person's name).** On the other hand, Siri will take up the command and confirm your demand by asking **Are you ready to send it?** At this junction, you can either say yes/no, or you can introduce a comment to the message using the input field and then hit Send.

When it is content that can't be shared directly, such as a weather forecast, Siri will screenshot the forecast and send it instead. Just say **Share this with [person's name],** Siri will get hold of the screenshot, and then verify the request with you in the same manner. Are **you ready to send it?** Before sending it. This Siri feature is compatible with photos, Apple News, Apple Podcasts, Safari web pages, Messages, Apple Music, Maps, and much more.

Asking Siri to Control Your HomeKit Devices

You can actually use Siri to control HomeKit devices at certain times

If you desire to put on the security lights at eight o'clock the later in the evening, simply tell the virtual assistant, Hey Siri, turn on the security at 8 p.m. The virtual assistant Siri also understands geolocation requests. For instance I could say Hey Siri, lock the doors when I leave.

Siri will set up an automation in the Automation tab of the Home app when prompted to control a HomeKit accessories in this manner. To delete the automation set up by Siri in the Home app, just swipe left over the automation and then click on the Delete link. For Siri commands to work with third-party devices you will need HomePod to convey the requests through. With the virtual assistant integration, third-party HomeKit accessories can be controlled with Siri commands to perform tasks such as broadcasting messages, creating reminders, managing devices, etc.

Enabling Siri Announce to Your Notifications

With iOS/iPadOS 15 Siri can herald notifications to you when you're putting on connected Beats headphones or AirPods. Before now, Siri announces message alerts however in iOS/iPadOS 15 the feature now spreads across to all notifications. So Siri will now automatically announce Time Sensitive notifications from apps when AirPods/Beats are connected if turned on.

You now decide which apps that aren't time sensitive get selected for Siri's announcement notifications functionality.

To enable announce notifications feature;

- Go to the **Settings** app on your device.

- Select **Notifications**.
- Beneath **Siri** choose **Announce Notifications**.
- Swipe the switch next to **Announce Notifications to** ON position. Green indicates it is enabled.
- Under the heading **Announce Notifications From** tap the switch of the apps you want Siri to announce to on.

If the alert announced by Siri supports activity, Siri will wait for your reply after reading the alert, so you don't need to declare "Hey Siri" to swing into action.

Sharing Your Screen on a FaceTime

FaceTime saw significant breakthroughs in iPadOS/iOS 15, and one of them is the SharePlay functionality. Apart from using FaceTime for calls, you can now share your screen with other individuals on a video call. This functionality also offers the means to stream TV shows or movies jointly with other people. Every person on FaceTime call with you will see the same synced video and playback controls. To access this feature;

- Go to the **FaceTime** app on your device.
- Click on the **New FaceTime** tab.
- Select the contacts that you wish to share your screen with, and the click **FaceTime** switch. Or

you can tap a current contact to begin a video call.

- Toggle the SharePlay icon at the upper-right of the interface in the new control panel when the call is secured.
- Click **Share My Screen** in the dropdown and after a three-second countdown, screen sharing will start.

When FaceTime screen sharing is active, you can switch to any app you want to share with the group or family member. You will see the sharing icon at upper-left of the display to signify that FaceTime screen sharing is on. Click on the icon to access the FaceTime control panel.

Apart from watching movies or TV shows with members, you can also use this feature to flick through a photo album.

While on FaceTime call you can swipe away the active caller's face to access extra screen space, and simply swipe them back into the view. When you're watching a display shared by another individual, the person's name will be displayed beneath the sharing icon with buttons to share the content with another person, send a message to the host, or love the content. Also via the SharePlay window you can view TV shows and movies or listen to songs together.

Sorry for this news, you will have to wait for the not too distant future iOS/iPadOS 15 and MacOS updates to explore this feature as Apple has was discontinued SharePlay in this updates.

Customizing iCloud Private Relay IP Address Settings

In iOS/iPadOS 15, Apple brought in iCloud Private Relay and it is a functionality of iCloud+ paid plans. It is made to encrypt every of the outgoing traffic on your device to prevent them from being accessed or blocked by someone.

Private Relay functions by transmitting web traffic to a server that is sustained by Apple to take away the IP address. Once the IP Address details have been erased, Apple transmits the traffic to another server sustained by a third-party firm that gives a momentary IP address and then conveys the traffic to its destination. This will in turn prevent the use of your IP address, browsing history and location by any one. When iCloud Private Relay is activated, you can decide how the designated IP address utilizes geographical data to hide your true locality.

And if you still desire to get local information when you're surfing in Safari, the IP address can keep your general location without having to be precise or if you want the relay to be extremely private, the IP address can simply carry your country and time zone.

To customize the feature;

- Open the Settings app on your device.
- Select your name at the upper section of the Settings screen.
- Click **iCloud**.
- And then select **Private Relay**.
- Toggle IP Address Location.
- And then you go for **Maintain General Location** which is the default setting or you can select **Use Country and Time Zone** option.

Turning Off/On iCloud Private Relay

iCloud Private Relay is one of the new features available in iCloud+ paid subscription and it is designed to encrypt every traffic going out of your device, making it unreadable and secure.

To toggle on or off Private Relay;

- Open the **Settings** app on your device.
- Toggle your name at the upper area of the Settings screen.
- Then click **iCloud**.
- Select **Private Relay**.
- Then swipe switch next to **Private Relay** On or Off. Green will indicate it is on.

Now that **Private Relay** is activated, via the **IP Address Location** link you can employ the option **Maintain General Location** which is the default setting to hold onto local information in browsing. Or

136

you can switch to the smaller geographically specific and additional private **Use Country and Time Zone** option.

Customizing the Address/Search Bar Design of Safari

With iOS/iPadOS 15 running in your device, you will get an all-new Safari design, which positions the address and search bar at the bottom of the browser interface by default. However, this default setting can be switched. You can switch back to the typical Safari design that has the address and search bar at the top of the display.

Switch the address and search bar design;

- Launch Safari on your device.
- By the left side of the Address/Search bar, toggle the aA icon.
- From a panel that will come up select the option Show Top Address Bar. And if you want to return to the bottom Address/Search bar design
- Click aA icon on address/search bar
- This time go for the option Show Bottom Tab Bar.
-

How to Add Text Size Functionality to Control Center.

To be able to use the Text Size feature on apps you have to bring in the functionality into the control panel first. To do this;

- Go to the **Settings** app on your device.
- Select **Control Center**.
- Click on the Plus sign + next to Text Size icon **AA**.

So, you can now access the Text Size feature from your device's Control Center or panel.

How to Customize Text Size

There might be some apps or settings that have text which are too small or too big. But with iOS 15/iPadOS 15 you can adjust the text size in the apps. Changing the Text Size in an app – the Text Size functionality works seamlessly in various apps, however several third-party apps do not support this functionality.

- Open the app you desire.
- Based on your device swipe up from the bottom of the screen, or swipe down from the top right corner to access the Control Center.
- Click on Text Size icon AA.
- Toggle on the app-only button at the button left corner.

138

- Select the text size you desire. Note by default text size is 100%, however you can reduce it to minimum 80% normal size or take it to the highest point 310%.

How to Access Sleep Trends in the Health app

iOS/iPadOS 15 comes with a Health app that features a new Trends functionality that enables you to see the trending patterns of your sleep, concluding that you put your Apple Watch to bed from onset.

To view be able your Sleep Trends;

- Open the Health app on your device.
- Navigate to **View Health Trends** under the tab **Trends**.
- And then toggle on **View Health Trends**.
- Navigate to the bottom of the interface until you see the trends associated with your sleep health.
- Click on a **trend** to see more about it, like expansion of the trend.
- Toggle on the option **Show More Data** to get additional information.

Health Trends establish the standard for you based on the information the Health app extracts from your Apple Watch and other apps for specific places. The Health app will prompt you when it detects low or high changes. However, when the data is consistent the data will be placed at the Not Trending tab of the Health Trends as a Daily Average.

We have various trends that you may see concerning your Sleep health and they include Respiratory Rate, Heart Rate and Sleep Duration.

Respiratory Rate – one of the new functionality in watchOS 8 is the capacity to keep tabs on respiratory rate. It is typically how many breaths per minute you draw during sleep. Normally your respiratory rate while sleeping should moderately stay steady over time, making it not show up in the **Trending** segment. However, if it's trending, whether low or high, then that might indicate health problems like Bradypnea, sleep apnea.

Sleep Duration is how long you spend sleeping every night and you can get data like average time asleep, in bed and your sleep goal.

Heart Rate determines how many beats per minute (bpm) your heart is performing during sleep. There is little distinction between Heart rate and Resting Heart Rate, the latter is the average heart beat per minute while relaxing for some minutes or when you're non-functioning.

Recording a FaceTime call on Your iPhone Via Your Mac

Note that you cannot locally record a FaceTime call via your iPhone or iPad screen but a Mac must be introduced to aid the process.

So to record a FaceTime call on your iPad or iPhone via a Mac;

- Use a 30-pin connector or lightning cable to link your iPad or iPhone.
- Launch QuickTime on the Mac through the app folder or dock. Select File from the Menu bar.
- And toggle on the option New Movie Recording.
- Tap the drop-down arrow next to the record button in the QuickTime interface.
- From the options, select your iPhone camera.
- Once you unlock the iPhone its screen will pop up on QuickTime interface on the Mac.
- Ensure to enable the volume bar on the Mac's QuickTime to get the audio of the call.
- Now launch the FaceTime app on iPad or iPhone.
- Toggle the Record button in QuickTime on your Mac.
- And initiate FaceTime calls on the iPad or iPhone.

- Tap the Stop button in QuickTime to end the recording when you're done with the conversation.
- Go to File in the Menu bar.
- Select the option **Save** from displayed options. Enter a name for recording as you desire.
- Select the location desire to store the recording.
- And go ahead to hit **Save** again. You can decide to save the file in Movies, Desktop, Downloads, etc.

Customizing Widgets in iPadOS 15

In iPadOS 15 you can subject your widgets to several customization ranging from the content they display to their size.

To customize a widget, **Long-press** on a Widget and tap the option **Edit Widget** from the small screen that shows up. Then you can do the changes you want on the Widget.

To remove a widget from the Home Screen of your device, select the option **Remove Widget** from the small screen that will be displayed when you tap and hold on the Widget. Or you can long-press on a clear space on the Home Screen to trigger the **jiggle mode**. Click the minus icon – at left top to remove the widget from the Home Screen.

Creating Smart Stack Widgets

The placing or piling of one widget over another is called a stack. Smart Stacks helps to display vital widgets automatically at the appropriate time and this is achieved via Widget Suggestions and Smart Rotate. Place a widget over another to on the Home screen to create a widget stack. And to bring in a preset smart stack;

- Long-press on a clear your Home Screen.
- On the jiggle mode click on the + icon at the top left corner.
- Click Smart Stack on the menus on the left side.
- Browse through the widgets to select the one you desire.
- Toggle the option Add Widget. Alternatively, you can decide to drag and drop it.

About the Author

Paul Spurgeon is a tech enthusiast with over 15 years' experience in the ICT sector. He is ardent of tech and passionately follows the latest technical and technological trends. His strength lies in figuring out the solution to sophisticated tech problems base on his technical know-how. Paul holds a Bachelor and a Master's Degree in Computer Science from Stanford University in Stanford, California.

Printed in Great Britain
by Amazon

67205181R00088